Greenhouse Gas Emission and Mitigation in Municipal Wastewater Treatment Plants

Greenhouse Gas Emission and Mitigation in Municipal Wastewater Treatment Plants

Xinmin Zhan, Zhenhu Hu, Guangxue Wu

IWA PUBLISHING

Published by **IWA Publishing**
Alliance House
12 Caxton Street
London SW1H 0QS, UK
Telephone: +44 (0)20 7654 5500
Fax: +44 (0)20 7654 5555
Email: publications@iwap.co.uk
Web: www.iwapublishing.com

First published 2018

British Library Cataloguing in Publication Data
A CIP catalogue record for this book is available from the British Library

ISBN: 9781780406305 (Paperback)
ISBN: 9781780406312 (eBook)
ISBN: 9781780409054 (ePUB)

Contents

Chapter 3

Chapter 4

Preface

Management and treatment of wastewater has always been an important part of civilisation. However, the development of wastewater treatment technologies is sometimes well behind economic development because of our slow understanding of the environment, ecology and health. Throughout history, drainage systems have been used to deliver wastewater from residential communities to the rivers, lakes and the sea, but the realisation of the importance of wastewater treatment had not been realised until the 19th century in England. At the beginning of the last century, the activated sludge process was invented, which has become the foundation of many wastewater treatment technologies being applied in municipal wastewater treatment plants (WWTPs). Since the start of this century, most countries have imposed stringent regulations and required tertiary treatment with efficient removals of organic matter, nitrogen and phosphorus from wastewater.

Effective wastewater treatment not only protects our local environment and ecology from serious damage but also protects human health. However, some concerns relating to wastewater treatment have been recently raised. One is that municipal WWTPs might be sources of greenhouse gas (GHG) emissions. During organic matter degradation through aerobic oxidation or anaerobic digestion, and nitrogen removal through nitrification and denitrification, GHGs, among which the three most potent ones are carbon dioxide (CO_2), nitrous oxide (N_2O) and methane (CH_4), are generated. In addition, indirect emissions of these gases, as a result of the inputs of chemicals and energy into WWTPs to the atmosphere, contribute to the carbon footprints of the municipal WWTPs. These emissions consequently impact the global environment and cause climate change: the rising of sea levels, the diminishing of ice in the Arctic Ocean and the Antarctic continent, the retreating of glaciers, and frequent extreme flood and draught events. This creates a paradox: in our efforts to improve the local water environment, have we sacrificed the global environment?

The contributions of WWTPs' GHG emissions to the national GHG inventories are relatively low, but they are expected to increase. One reason is that many countries have now imposed stringent wastewater discharge limits on their WWTPs. Municipal WWTPs are developing and applying novel technologies and inputting more materials and energy to achieve higher discharge water quality. These will bring about higher direct and indirect GHG emissions. The other reason is the increasing population that now have centralized wastewater treatment services. The increasing protein contents in our diets, which cause higher nitrogen concentration in wastewater, may also lead to higher N_2O emissions from WWTPs.

This book provides an overview of the mechanisms and pathways of GHGs emissions from WWTPs and explores potential technical and managerial methods to mitigate. Its chapters examine biochemical reactions generating GHGs in wastewater treatment unit processes, direct and indirect GHG emissions and emission factors, Intergovernmental Panel on Climate Change (IPCC) framework for calculating national GHG emission inventories for the wastewater sector, monitoring of direct GHG emissions, CH_4 and N_2O generation and mitigation in wastewater treatment facilities, and life cycle assessment of municipal WWTPs. This book will hopefully be a valuable resource for environmental engineering and science undergraduate and postgraduate students, for researchers working in relevant areas, water professionals and government policy makers.

I would like to thank my co-authors for their contributions and in particular, researchers in my group, such as Qingfeng Yang, Li Shi and Yan Yang, who have helped me in improving the quality of the pictures and graphs. I would also like to thank the editors of IWA Publishing, Maggie Smith, Mark Hammond and Niall Cunniffe, who were always helpful.

Finally, I would like to share a dictum of an ancient Chinese philosopher Lao Zi with our readers, which might answer the question posted above, "Harmony between Nature and Human Beings".

Xinmin Zhan
College of Engineering and Informatics
National University of Ireland, Galway
Ireland

Chapter 1

Warming climate and greenhouse gases

Xinmin Zhan, Guangxue Wu and Zhenhu Hu

1.1 WARMING CLIMATE

Climate change and global warming are two well-known terms all over the world, among the industry, academia, farmers, policymakers and the general public. The warming of our climate is evidenced by rising global temperatures, rising sea levels, diminishing ices in the Arctic Ocean and the Antarctic continent, retreating glaciers, and frequent extreme flood and drought events (NASA [National Aeronautics & Space Administration], 2017). The globally average temperature (combining the land and ocean surfaces) has had an increasing trend since 1880, with an increase of 0.85°C over the period of 1880–2012 (IPCC [Intergovernmental Panel on Climate Change], 2014). Even though we have implemented a number of actions to fight global warming, it is clear that the warming trend is continuing, as 15 of the 16 warmest years on record have taken place since 2001, and the year 2015 had a temperature 1°C or more higher the 1880–1899 average (NASA, 2017).

Oceans are always regarded as the conditioner maintaining the global temperature constant. Unfortunately, the ocean surface temperature also shows a warming tendency with the upper 75 m warmer by 0.11°C (0.05–0.13°C) per decade in the period 1971–2010 (IPCC, 2014). The upper 700 m of ocean water warmed from 1971 to 2010 (95–100% probability confidence) (IPCC, 2014) and the global oceanic temperature in the upper 700 m rose by 0.15°C between 1955 and 2016 (NODC [National Oceanographic Data Center], 2017).

The warming effects vary by region, with the effects being more obvious and significant in some regions than others (Figure 1.1). In Ireland, the observational

 Greenhouse Gas Emission and Mitigation in Municipal Wastewater Treatment Plants
Xinmin Zhan, Zhenhu Hu, Guangxue Wu
doi: 10.2166/9781780406312_001

data from five long stations show that the mean annual temperature increased by approximately 0.8°C over the period 1900–2012 (Gleeson *et al.* 2013).

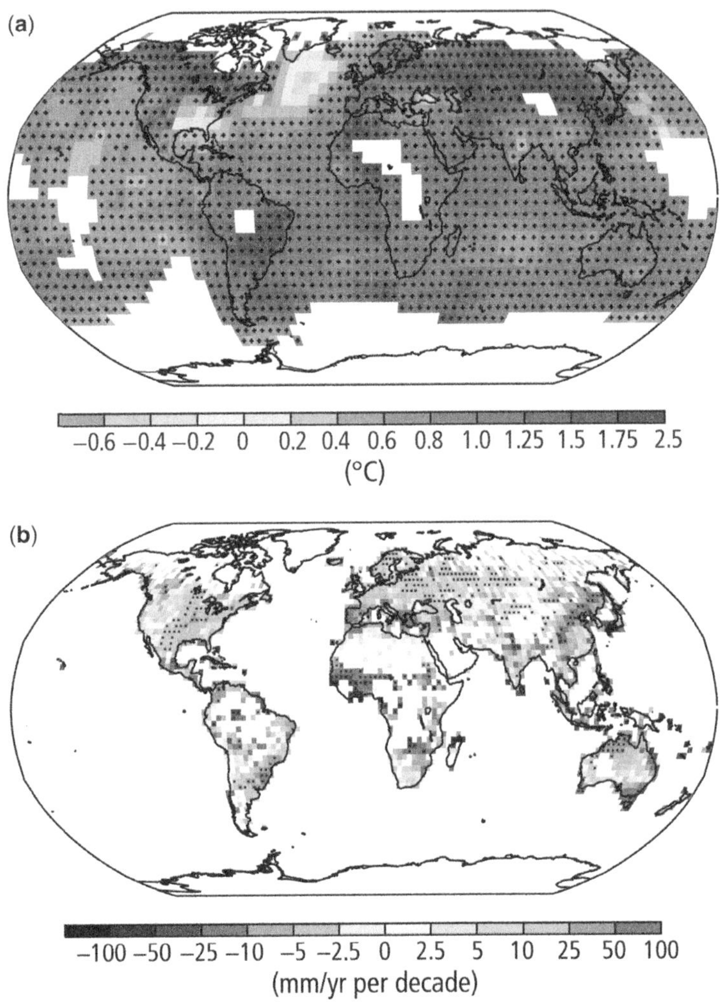

Figure 1.1 An uneven changing global climate system ((a) observed surface temperature change from 1901 to 2012; (b) observed precipitation change from 1951 to 2010) (IPCC, 2014).

Intensive scientific research has shown that the cause of the warming climate is anthropogenic greenhouse gas emissions, which have increased significantly since the pre-industrial era (IPCC, 2014): there is a 95–100% probability that the anthropogenic increase in GHG emissions has led to the global average surface temperature increase from 1951 to 2010. Between 1750 and 2011, 2040 ± 310 Gt $CO_{2\text{-eq}}$ (1 Gt = 10^9 tonnes) was emitted to the atmosphere, half of which occurred in the last 40 years, with anthropogenic GHG emissions in 2010 reaching 49 ± 4.5 Gt $CO_{2\text{-eq}}$/year. Currently, the atmospheric CO_2 level is 407 parts per million, while it was 280 parts per million 250 years ago, in the mid-18th century (NASA, 2017). Total annual anthropogenic GHG emissions increased sharply after 2000, by 1.0 Gt $CO_{2\text{-eq}}$ per year from 2000 to 2010. The annual increase was 0.4 $CO_{2\text{-eq}}$ per year over the period of 1970 to 2000 (Figure 1.2). This was due to the rapid economic development in emerging and developing countries, for example China, India, and Brazil.

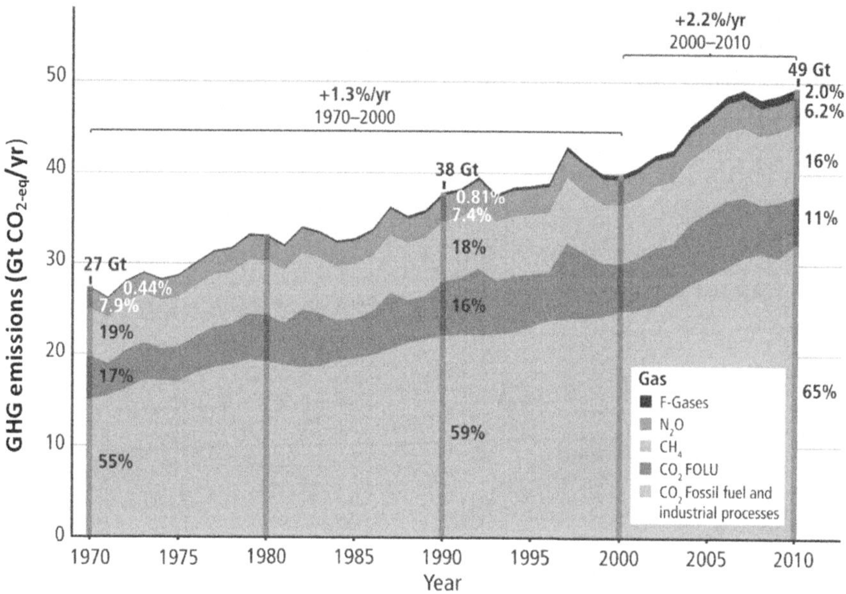

Figure 1.2 Total annual anthropogenic GHG emissions by gases over the period of 1970–2010 (IPCC, 2014). *Note*: F-Gases are fluorinated gases covered under the Kyoto Protocol, and, CO_2 FOLU is CO_2 emission from the Forestry and Other Land Use category (FOLU).

An increasingly warming climate will damage our natural environment and human living systems. This damage will vary from region to region but will include: loss of livelihoods, infrastructure and economic stability; increased flood and drought events; increased heat-related human mortality; food-shortage and lack of food security; and reduced biodiversity of fauna and flora. Hence, all the states in the world are working together to fight the warming climate with the goal of limiting global temperature rise in the year 2100 to within 2°C of the pre-industrial temperature. In order to achieve this target, global anthropogenic GHG emissions need to be reduced by 40–70% by 2050 compared to the 2010 level, and reduced by almost 100% by 2100 (IPCC, 2014).

The three major GHG gases are carbon dioxide (CO_2), methane (CH_4) and nitrous oxide (N_2O). They accounted for 76%, 16% and 6.2%, respectively, of total anthropogenic GHG emissions in 2010 (IPCC, 2014).

1.2 GREENHOUSE GASES AND GLOBAL WARMING POTENTIAL

Without greenhouse gases, after solar energy arrived at the Earth's surface, the heat would be radiated back into space. However, due to the presence of these gases which are spread over the Earth's surface and able to absorb the outgoing heat, some heat is trapped between the Earth's surface and the low atmosphere (Figure 1.3). This extra heat causes the temperature of the Earth's surface to rise.

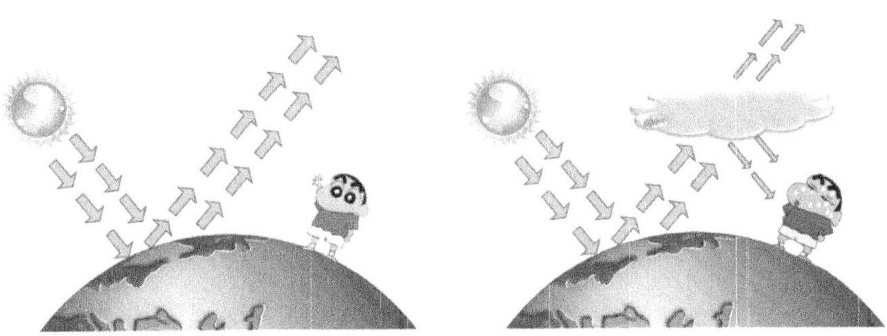

Figure 1.3 Cause of global warming.

In order to compare the impacts of different GHG gases on global warming, a metric, global warming potential (GWP), has been developed. GWP is defined as the amount of energy which the emission of 1 tonne of a gas will absorb over a certain period of time, normally 100 years, relative to the emission of 1 tonne of CO_2. Hence, the GWP value of CO_2 is 1 in 100 years, while the GWPs of CH_4 and N_2O are 28 and 265 over 100 years, respectively. Table 1.1 lists the GWP values for common greenhouse gases (IPCC, 2014). It should be noted that GWPs vary according to the methodology defining the impacts of future warming on the carbon cycle and the calculation methods used (USEPA [United States Environmental Protection Agency], 2017).

Table 1.1 GWP values for greenhouse gases.

	GWPs	
	Cumulative Forcing Over 100 Years	**Cumulative Forcing Over 20 Years**
CO_2	1	1
CH_4	28	84
N_2O	265	264
CF_4	6630	4880
HFC-152	138	506

Source: IPCC (2014).

Based on GWP, we can use the following two equations to convert the emissions of CH_4 and N_2O to the equivalent emissions of CO_2:

$$CO_{2\text{-eq}} = 28 \times M_{CH_4} \tag{1.1}$$

$$CO_{2\text{-eq}} = 265 \times M_{N_2O} \tag{1.2}$$

where, M_{CH_4} and M_{N_2O} are the emissions of CH_4 and N_2O gases respectively (tonne); and $CO_{2\text{-eq}}$ is the equivalent emission of CO_2 (tonne).

Anthropogenic CO_2 emissions are mainly due to burning of fossil fuels, and deforestation. Anthropogenic N_2O emissions are due to agricultural use of chemical fertilizers, land-spreading of animal manure, and burning of biomass and fossil fuels. Anthropogenic CH_4 emissions are caused by decomposition of organic matter during human activities, for example, land-spreading and storage of animal manure, rumen fermentation of ruminant livestock, and gas loss from anaerobic digesters of biowastes. They are also produced in wastewater treatment plants (WWTPs). CO_2, CH_4, and N_2O are the main greenhouse gases generated during the wastewater treatment processes (Hendrickson *et al.* 2015).

1.3 GENERATION OF CO_2 IN WASTEWATER TREATMENT PLANTS

CO_2 is the largest contributor to greenhouse gas emissions. It is the product of the oxidation of organic matter:

$$C + O_2 \rightarrow CO_2 \tag{1.3}$$

In the wastewater treatment process, CO_2 is produced due to the oxidation of organic matter by microorganisms in aerobic or anaerobic conditions.

In aerobic conditions, microorganisms break down the organic matter so as to obtain energy and synthesize new cells through which organic matter is removed from wastewater:

$$\text{Organic matter} + O_2 + \text{nutrients (nitrogen, phosphorus, etc.)} \rightarrow \text{new cells} + CO_2 + H_2O + \cdots \quad (1.4)$$

If we assume that organic matter is biologically decomposed and nitrification is ignored, and use $C_aH_bO_cN_d$ to represent organic matter, and $C_wH_xO_yN_z$ to represent the growth of new microorganism biomass, the equation above is re-written as:

$$C_aH_bO_cN_d + 0.5(ny + 2s + r - c)O_2 + (d - nz)H^+ \rightarrow nC_wH_xO_yN_z + sCO_2 + rH_2O + (d - nz)NH_4^+ \quad (1.5)$$

The chemical composition of organic matter in wastewater can be represented approximately by the formula: $C_{18}H_{19}O_9N$ (Henze *et al.* 1996). If the growth of biomass and nitrification is ignored, then Eq. 1.5 is simplified to:

$$C_{18}H_{19}O_9N + 17.5O_2 + H^+ \rightarrow 18CO_2 + 8H_2O + NH_4^+ \quad (1.6)$$

In anaerobic conditions, when microorganisms degrade organic contaminants, CO_2 is also produced with CH_4, which will be detailed in a later section.

If chemical processes are adopted in WWTPs to breakdown organic matter, CO_2 is also produced (Tchobanoglous *et al.* 2003):

$$\text{organic matter} + \text{oxidants } (Cl, O_3, H_2O_2, \text{etc.}) \rightarrow \text{intermediate oxygenated molecules} \quad (1.7)$$

$$\text{intermediate oxygenated molecules} + \text{oxidants } (Cl, O_3, H_2O_2, \text{etc.}) \rightarrow \text{simple end products } (CO_2, H_2O, \text{etc.}) \quad (1.8)$$

Since chemical oxidation processes are not commonly used in municipal WWTPs, we only need to consider CO_2 production and emission in aerobic and anaerobic biological wastewater treatment processes. The IPCC excludes CO_2 emission in the greenhouse gas emission inventory for WWTPs because the IPCC recognizes that CO_2 emission from WWTPs is from biogenic origins and it doesn't contribute to the increase of anthropogenic CO_2 concentrations in the atmosphere (Shaw & Koh, 2014). However, organic contaminants with fossil carbon origins, like surfactants, pharmaceuticals and personal care products, exist in wastewater influent into wastewater treatment plants, and account for up to 8–14% of total chemical oxygen demand (COD) concentration in a combined flow of residential and industrial wastewater, and 4–7% of COD in residential wastewater (Law *et al.*

2013). Hence, the CO_2 emissions due to the biodegradation of these fossil origin organic contaminants should be taken into account when quantifying WWTP GHG emissions.

Indirectly, carbon dioxide is emitted in wastewater treatment processes as a result of fossil fuel combustion to meet the large energy requirement of WWTPs (Ashrafi *et al.* 2014).

1.4 GENERATION OF CH_4 IN WASTEWATER TREATMENT PLANTS

CH_4 is the second-most potent greenhouse gas after CO_2 in terms of impacts on global warming. Its GWP is 28 times over a 100-year basis. The concentration of CH_4 in the atmosphere has increased from 0.7 parts per million in pre-industrial times to 1.8 parts per million today (Zeebe, 2013). The trend still continues, with atmospheric CH_4 concentrations increasing by approximately 7 parts per billion per year (Vollmer *et al.* 2015). Therefore, the mitigation of methane emission is very important when tackling global warming. Methane generation by anaerobic respiration has both natural and anthropogenic sources.

CH_4 is generated by microorganisms under anaerobic conditions in sewers and WWTPs when they decompose organic matter:

$$\text{Organic matter} + H_2O + \text{nutrients (nitrogen, phosphorus, etc.)} \rightarrow \\ \text{new cells} + \text{resistant organic matter} + CO_2 + CH_4 + NH_3 + H_2S + \cdots \quad (1.9)$$

The biological conversion of organic matter in wastewater to CH_4 is complex and involves a series of microbial metabolic stages which occur in anaerobic conditions (Weiland, 2010). The conversion process consists of four steps: hydrolysis, acidogenesis, acetogenesis, and methanogenesis. Each step is completed by specific types of microorganisms on substrates generated from the preceding steps. This is illustrated in Figure 1.4.

Organic matter in wastewater is composed of carbohydrates (which include cellulose, hemicellulose and lignin), proteins, lipids, and other natural or synthetic compounds. Hydrolysis is the first step for the anaerobic degradation of organic matter (Hu *et al.* 2008). Via enzymes excreted by facultative and obligate bacteria and fungi, these organic compounds are hydrolyzed into smaller molecules, such as amino acids, fatty acids, peptides, and CO_2, as well as a small amount of acetic acid and H_2 (Cheng, 2009).

The second step is acidogenesis. The amino acids, fatty acids, polysaccharides, and peptides generated in the hydrolysis step are further converted to volatile fatty acids (VFAs) by acidogenic bacteria. Small amounts of acetic acid, H_2, and CO_2 are also produced.

The produced non-acetic acid VFAs are then converted into acetic acid, H_2, and CO_2. This step is acetogenesis. Acetic acid, H_2 and CO_2 are precursors for methane generation in the last step – methanogenesis. Acetic acid and H_2/CO_2 produced in the acetogenesis and previous steps are utilized by archaea to synthetize CH_4. There are two different biochemical pathways for CH_4 generation based on the substrates used. If acetic acid is used as the substrate, the reaction is as described in Eq. 1.10:

$$CH_3COOH \rightarrow CH_4 + CO_2 \tag{1.10}$$

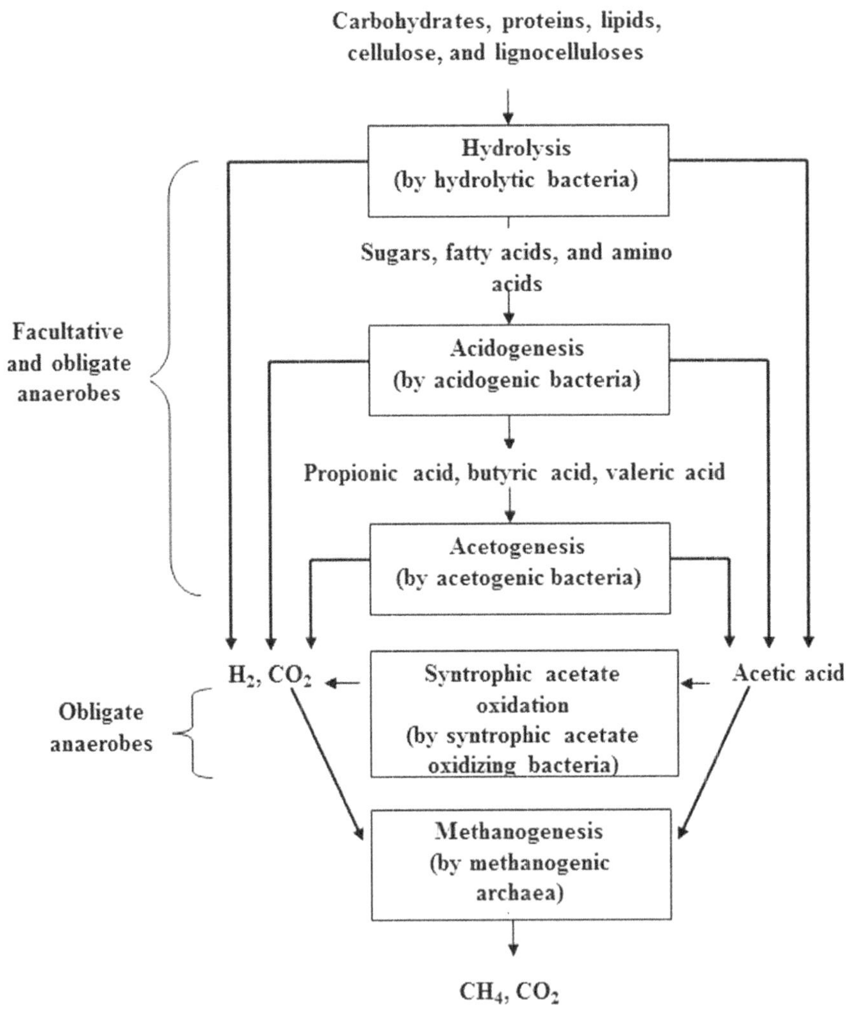

Figure 1.4 Schematic diagram of biological methane production from organic compounds (Cheng, 2009; Wang *et al.* 2015; Westerholm *et al.* 2016).

This pathway, known as acetotrophic methanogenesis, is completed by obligate or facultative acetate-utilizing archaea. Acetotrophic methanogenesis is widely recognized as the main pathway for CH_4 production. H_2 and CO_2 can also be used as substrates to form CH_4, as described in Eq. 1.11:

$$4H_2 + CO_2 \rightarrow CH_4 + 2H_2O \tag{1.11}$$

This pathway, hydrogenotrophic methanogenesis, is completed by obligate or facultative hydrogen-utilizing archaea.

Eq. 1.9 is simplified as:

$$\begin{aligned} C_aH_bO_cN_d + (d - nx)H^+ &\rightarrow nC_wH_xO_yN_z + mCH_4 + sCO_2 \\ &+ rH_2O + (d - nx)NH_4^+ \end{aligned} \tag{1.12}$$

where s = a − nw − m, and r = c − ny − 2s.

If the anaerobic degradation process is complete, then:

$$\begin{aligned} &C_aH_bO_cN_d + \frac{4a - b - 2c + 3d}{4}H_2O + dH^+ \\ &\rightarrow \frac{4a + b - 2c - 3d}{8}CH_4 + \frac{4a - b + 2c + 3d}{8}CO_2 + dNH_4^+ \end{aligned} \tag{1.13}$$

Based on Eq. 1.13, if we know the composition of the organic compounds and their amounts, the theoretical yield of CH_4 from organic compounds can be predicted. If $C_{18}H_{19}O_9N$ is used to describe the chemical composition of municipal wastewater, then:

$$C_{18}H_{19}O_9N + 9.5H_2O + H^+ \rightarrow 8.75CH_4 + 9.25CO_2 + NH_4^+ \tag{1.14}$$

when d = 0, namely when nitrogen is ignorable, then Eq. 1.13 is simplified to the well-known Buswell equation (Horbaj, 2004):

$$C_aH_bO_c + \left(a - \frac{b}{4} - \frac{c}{2}\right)H_2O \rightarrow \left(\frac{a}{2} + \frac{b}{8} - \frac{c}{4}\right)CH_4 + \left(\frac{a}{2} - \frac{b}{8} + \frac{c}{4}\right)CO_2 \tag{1.15}$$

1.5 N_2O GENERATION IN WASTEWATER TREATMENT FACILITIES

N_2O can be emitted during both nitrification and denitrification. The lifetime of N_2O is about 120 years, which causes a long-term effect on global warming. Its GWP is 265 in 100 years. The atmospheric N_2O concentration has been increasing at 0.73 ± 0.03 parts per billion/year over the last three decades (IPCC, 2014) and, at present, it is approximately 330 parts per billion (IPCC, 2014). Because CH_4 and N_2O have much shorter lifetimes than that of CO_2, it is good practice to mitigate global warming by reducing non-CO_2 emissions (Montzka *et al.* 2011). Nowadays, N_2O emission receives extensive attention due to its high GWP, and should be carefully controlled from all sectors, including agricultural farming and wastewater management systems. Apart from GHG impacts, N_2O in the atmosphere reacts with atomic oxygen to form NO, which induces the destruction of the stratospheric ozone (Ravishankara *et al.* 2009).

In the EU-27, N_2O is mainly emitted in waste treatment activities (10 million tonnes $CO_{2\text{-eq}}$ emission), nitric acid production (27 million tonnes $CO_{2\text{-eq}}$ emission), and livestock farming (21 million tonnes $CO_{2\text{-eq}}$ emission) (López *et al.* 2013).

Biological nitrogen removal (BNR) in WWTPs includes nitrification and denitrification. More complicated than generation of CO_2 and CH_4 in WWTPs, during BNR, typically, there are three main pathways for N_2O emission (Figures 1.5 and 1.6), i.e., nitrifier denitrification, hydroxylamine (NH_2OH) oxidation and heterotrophic denitrification. Other pathways may also contribute to N_2O emission. For instance, chemidenitrification and the nonenzymatic decomposition of nitrite can also lead to the production of N_2O when there is a high concentration of nitrite, NO_2^-, in wastewater treatment systems (Anderson & Levine, 1986).

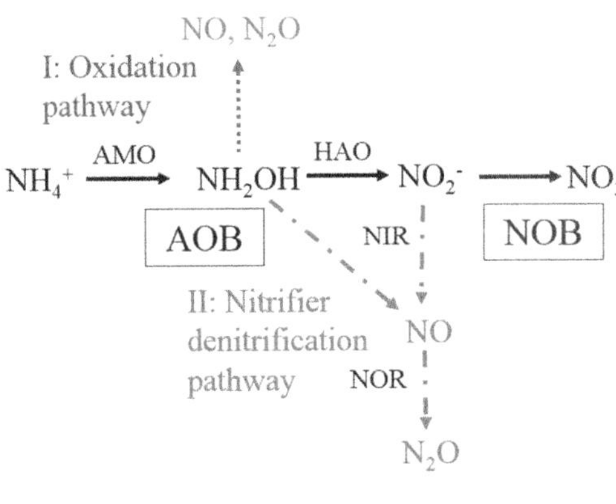

Figure 1.5 Processes which occur during nitrification with possible N_2O emission pathways. AMO: ammonia monooxygenase; HAO: hydroxylamine dehydrogenase.

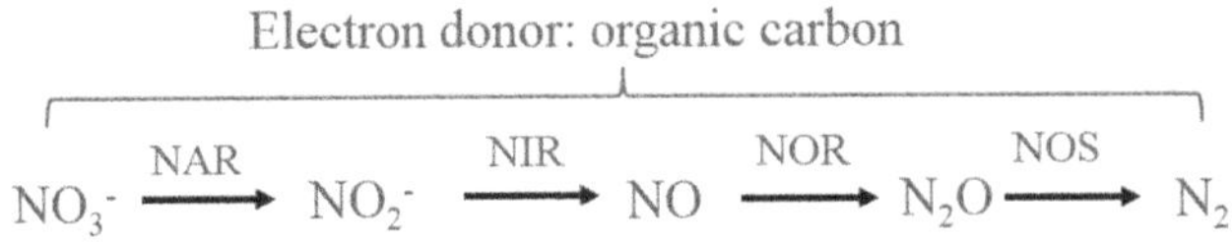

Figure 1.6 Processes which occur during denitrification with possible N_2O production.

1.5.1 N_2O emission during nitrification

Complete oxidation of ammonium (NH_4^+) to nitrate (NO_3^-) consists of two successive steps: oxidation of NH_4^+ to NO_2^-; and oxidation of NO_2^- to NO_3^-. Each step is conducted by a specific group of nitrifiers. The first step is achieved by ammonia oxidizing bacteria (AOB) and the second step is by nitrite oxidizing bacteria (NOB). AOB first nitrify NH_4^+ to NH_2OH, which is catalyzed by an NH_4^+ monooxygenase, and then further nitrify NH_2OH to NO_2^-, which is catalyzed by a hydroxylamine oxidoreductase. N_2O emission during these steps is due to either chemical oxidation of NH_2OH or nitrifier denitrification of NO_2^- (Figure 1.5), however, the contributions of these two pathways to N_2O emission during nitrification are not clear.

Nitrifier denitrification through which NO_2^- is reduced sequentially to NO and N_2O is mainly induced by a low oxygen concentration, which limits further oxidation of NO_2^- to NO_3^-, and consequently the accumulation of NO_2^-. Nitrifier denitrification might be used to protect AOB from toxic NO_2^- accumulation (Shaw *et al.* 2006). During nitrite nitrification, NO_2^- is the electron accepter, with the emission of N_2O as the intermediate similar to heterotrophic denitrification, while the electron donor may be organic matter or Fe or Cu (Wunderlin *et al.* 2012), NH_2OH (Ritchie & Nicholas, 1972; Hooper & Terry, 1979; Kim *et al.* 2010), and ammonia or H_2 (Bock *et al.* 1995). Some nitrifiers such as *Nitrosomonas europaea* have a copper-dependent nitrite reductase (NIR) and a NO reductase (NOR), but not a N_2O reductase (NOS) (Richardson *et al.* 2009), thereby producing N_2O rather than consuming it.

During NH_2OH oxidation, NH_2OH is oxidized to NO directly and then to N_2O. The production of N_2O via NH_2OH oxidation is likely to be minimal due to the low concentration of NH_2OH accumulation under normal operating conditions of wastewater treatment facilities (Wunderlin *et al.* 2012). NH_2OH would react with O_2 and NO_2^- chemically with the release of N_2O (Ritchie & Nicholas, 1972; Stüven *et al.* 1992; Schmidt *et al.* 2004; Wunderlin *et al.* 2012), with the hypothesized mechanisms as follows:

$$NH_2OH + 0.5O_2 \rightarrow 0.5N_2O + 1.5H_2O \qquad (1.16)$$

$$NH_2OH + NO_2^- + H^+ \rightarrow N_2O + 2H_2O \qquad (1.17)$$

The two reactions are also known as auto-oxidation of NH_2OH and chemo-oxidation of NH_2OH, respectively.

In complex nitrification systems, the above two processes – nitrifier denitrification and chemo-oxidation of NH_2OH – may occur simultaneously. N_2O emission is favored through NH_2OH oxidation at high NH_4^+ and low NO_2^- concentrations, while nitrifier denitrification by AOB is favored at higher NO_2^- and lower NH_4^+

concentrations (Wunderlin *et al.* 2012). Therefore, for mitigating N_2O emission, low NH_4^+ and NO_2^- concentrations should be maintained. Therefore, in nitrification wastewater treatment systems, the NH_2OH oxidation pathway would contribute more to the total N_2O emission than nitrifier denitrification during the initial state when NH_4^+ is high (for example 65% contribution of NH_2OH oxidation process) (Rathnayake *et al.* 2013), but in the later stage, because of the low NH_4^+ concentration, its contribution would be similar to nitrifier denitrification. N_2O emission can be used as an indicator of the failure of nitrification (Butler *et al.* 2009).

1.5.2 N_2O emission during denitrification

Wastewater treatment facilities always have biological denitrification zones where NO_3^- is reduced to N_2 by heterotrophic organisms, i.e., denitrifiers. A complete heterotrophic denitrification consists of sequential reductive steps from NO_3-N to NO_2-N, NO, N_2O and finally to N_2, as shown in Figure 1.6. Four different denitrification reductases, namely nitrate reductase (NAR), nitrite reductase (NIR), NO reductase (NOR) and N_2O reductase (NOS), catalyze the biochemical reactions for each step respectively (Zumft, 1997). Under optimal conditions, the reaction rates of each step keep up with each other, so there are no accumulations of intermediates – NO_2, NO and N_2O – and N_2 is the end product. However, if the operating condition is not optimal, unmatched reaction rates among all enzymes in the reduction metabolism pathway are likely to result in the accumulation of undesired intermediates such as N_2O. N_2O accumulates due to the imbalanced reaction rates between N_2O production and its reduction. Since NO_2^- and NO are the precursors of NO and N_2O generation, respectively, NO_2^- accumulation favors the formation of NO and N_2O. Among the four steps, the N_2O reduction is the slowest step (Cervantes *et al.* 2001) and NOS is prone to inhibition by high concentrations of NO_2^- (Gejlsbjerg *et al.* 1998).

During denitrification, NO_3-N reduction and NO_2-N reduction compete for electron donors (Van Rijn *et al.* 1996). The NO_2-N reduction rate is smaller than the NO_3-N reduction rate under low influent organic carbon/nitrogen (C/N) ratios. NO_2-N would be built up during denitrification when electron donors are insufficient and the NO_3-N reduction out-competes the NO_2-N reduction under limited electron supply conditions (Oh & Silverstein, 1999).

At a microcosmic level, many heterotrophs are able to contribute to denitrification, with some denitrifying NO_3^- to N_2 gas without NO_2^- accumulation (such as *Pseudomonas pseudoalcaligenes*), some denitrifying NO_3^- to N_2 gas with NO_2^- accumulation (such as *Bacillus niacini*), and some denitrifying NO_3^- to NO_2^- only (such as *Staphylococcus* sp.) (Martienssen & Schöps, 1999). Therefore, N_2O emission during denitrification is affected by the acclimated microbial communities. For example, denitrifiers such as glycogen accumulating organisms (GAOs) and denitrifiers utilizing internal organic carbons possess a high N_2O emission potential during denitrification. Finally, factors affecting denitrifying enzymes' activities may also induce N_2O accumulation and emission. For example, a low dissolved oxygen (DO) concentration may inhibit NOS enzymes more seriously than NIR or NOR, resulting in a high N_2O emission.

1.6 SUMMARY

CO_2, CH_4 and N_2O are the three major greenhouse gases leading to global warming. Their emissions in wastewater treatment facilities are due to the biochemical reactions occurring in these facilities to remove organic matter and nitrogen from wastewater: aerobic and anaerobic decomposition of organic matter, oxidation of NH_4^+ by autotrophic nitrifiers, and reduction of oxidized nitrogen (NO_2^- and NO_3^-) by heterotrophic denitrifiers. In order to control their emissions from wastewater treatment facilities, their emission mechanisms should be well understood.

1.7 REFERENCES

Anderson I. C. and Levine J. S. (1986). Relative rates of nitric oxide and nitrous oxide production by nitrifiers, denitrifiers, and nitrate respirers. *Applied and Environmental Microbiology*, **51**(5), 938–945.

Ashrafi O., Yerushalmi L. and Haghighat F. (2014). Greenhouse gas emission and energy consumption in wastewater treatment plants: impact of operating parameters. *Clean-Soil Air Water*, **42**(3), 207–220.

Bock E., Schmidt I., Stüven R. and Zart D. (1995). Nitrogen loss caused by denitrifying Nitrosomonas cells using ammonium or hydrogen as electron donors and nitrite as electron acceptor. *Archives of Microbiology*, **163**(1), 16–20.

Butler M., Wang Y., Cartmell E. and Stephenson T. (2009). Nitrous oxide emissions for early warning of biological nitrification failure in activated sludge. *Water Research*, **43**(5), 1265–1272.

Cervantes F. J., David A. and Gómez J. (2001). Nitrogen removal from wastewaters at low C/N ratios with ammonium and acetate as electron donors. *Bioresource Technology*, **79**(2), 165–170.

Cheng J. (2009). Biomass to Renewable Energy Processes. CRC Press, Boca Raton, FL.

Gejlsbjerg B., Frette L. and Westermann P. (1998). Dynamics of N_2O production from activated sludge. *Water Research*, **32**(7), 2113–2121.

Gleeson E., McGrath R. and Treanor M. (2013). Ireland's Climate: The Road Ahead. Met Éireann, Dublin.

Hendrickson T. P., Nguyen M. T., Sukardi M., Miot A., Horvath A. and Nelson K. L. (2015). Life-cycle energy use and greenhouse gas emissions of a building-scale wastewater treatment and nonpotable reuse system. *Advances in Environmental Science and Technology*, **49**(17), 10303–10311.

Henze M., Harremoes P., Arvin E. and Cour Jansen J. (1996). Wastewater Treatment. Lyngby, Springer, Berlin.

Hooper A. B. and Terry K. (1979). Hydroxylamine oxidoreductase of *Nitrosomonas*: production of nitric oxide from hydroxylamine. *Biochimica et Biophysica Acta (BBA)-Enzymology*, **571**(1), 12–20.

Horbaj P. (2004). Theoretical calculation of methane formation from municipal waste. *Chemicke Listy*, **98**(3), 137–141.

Hu Z. H., Liu S. Y., Yue Z. B., Yan L. F., Yang M. T. and Yu H. Q. (2008). Microscale analysis of in vitro anaerobic degradation of lignocellulosic wastes by rumen microorganisms. *Advances in Environmental Science and Technology*, **42**(1), 276–281.

IPCC (2014). Climate change 2014: synthesis report. In: Contribution of Working Groups I, II and III to the Fifth Assessment Report of the Intergovernmental Panel on Climate Change, Core Writing Team, R. K. Pachauri and L. A. Meyer (eds), IPCC, Geneva, Switzerland, pages 41, 46, 87.

Kim S.-W., Miyahara M., Fushinobu S., Wakagi T. and Shoun H. (2010). Nitrous oxide emission from nitrifying activated sludge dependent on denitrification by ammonia-oxidizing bacteria. *Bioresource Technology*, **101**(11), 3958–3963.

Law Y., Jacobsen G. E., Smith A. M., Yuan Z. and Lant P. (2013). Fossil organic carbon in wastewater and its fate in treatment plants. *Water Research*, **47**(14), 5270–5281.

López J. C., Quijano G., Souza T. S., Estrada J. M., Lebrero R. and Muñoz R. (2013). Biotechnologies for greenhouse gases (CH_4, N_2O, and CO_2) abatement: state of the art and challenges. *Applied Microbiology and Biotechnology*, **97**(6), 2277–2303.

Martienssen M. and Schöps R. (1999). Population dynamics of denitrifying bacteria in a model biocommunity. *Water Research*, **33**(3), 639–646.

Montzka S. A., Dlugokencky E. J. and Butler J. H. (2011). Non-CO_2 greenhouse gases and climate change. *Bionature*, **476**(7358), 43.

NASA (2017). NASA, NOAA Analyses Reveal Record-Shattering Global Warm Temperatures in 2015. NASA.

NODC (2017). Global Ocean Heat and Salt Content: Basin Time Series. Accessed via National Oceanographic Data Center website at https://www.nodc.noaa.gov/OC5/3M_HEAT_CONTENT/basin_avt_data.html

Oh J. and Silverstein J. (1999). Acetate limitation and nitrite accumulation during denitrification. *Journal of Environmental Engineering*, **125**(3), 234–242.

Rathnayake R., Song Y., Tumendelger A., Oshiki M., Ishii S., Satoh H., Toyoda S., Yoshida N. and Okabe S. (2013). Source identification of nitrous oxide on autotrophic partial nitrification in a granular sludge reactor. *Water Research*, **47**(19), 7078–7086.

Ravishankara A., Daniel J. S. and Portmann R. W. (2009). Nitrous oxide (N_2O): the dominant ozone-depleting substance emitted in the 21st century. *Acta Agriculturae Scandinavica Section A Animal Science*, **326**(5949), 123–125.

Richardson D., Felgate H., Watmough N., Thomson A. and Baggs E. (2009). Mitigating release of the potent greenhouse gas N_2O from the nitrogen cycle–could enzymic regulation hold the key? *Trends in Biotechnology*, **27**(7), 388–397.

Ritchie G. and Nicholas D. (1972). Identification of the sources of nitrous oxide produced by oxidative and reductive processes in *Nitrosomonas europaea*. *Biochemical Journal*, **126**(5), 1181–1191.

Schmidt I., van Spanning R. J. and Jetten M. S. (2004). Denitrification and ammonia oxidation by *Nitrosomonas europaea* wild-type, and NirK-and NorB-deficient mutants. *Microbiology*, **150**(12), 4107–4114.

Shaw A. R. and Koh S. H. (2014). Gaseous emissions from wastewater facilities. *Water Environment Research*, **86**(10), 1284–1296.

Shaw L. J., Nicol G. W., Smith Z., Fear J., Prosser J. I. and Baggs E. M. (2006). *Nitrosospira spp.* can produce nitrous oxide via a nitrifier denitrification pathway. *Environmental Microbiology*, **8**(2), 214–222.

Stüven R., Vollmer M. and Bock E. (1992). The impact of organic matter on nitric oxide formation by Nitrosomonas europaea. *Archives of Microbiology*, **158**(6), 439–443.

Tchobanoglous G., Burton F. L., Stensel H. D. and Metcalf and Eddy Inc. (2003). Wastewater Engineering: Treatment and Reuse. McGraw-Hill Education, New York.

USEPA (2017). Understanding Global Warming Potentials by USEPA. Available at https://www.epa.gov/ghgemissions/understanding-global-warming-potentials (accessed 10 April 2017).

Van Rijn J., Tal Y. and Barak Y. (1996). Influence of volatile fatty acids on nitrite accumulation by a *Pseudomonas stutzeri* strain isolated from a denitrifying fluidized bed reactor. *Applied and Environmental Microbiology*, **62**(7), 2615–2620.

Vollmer M. K., Rhee T. S., Rigby M., Hofstetter D., Hill M., Schoenenberger F. and Reimann S. (2015). Modern inhalation anesthetics: potent greenhouse gases in the global atmosphere. *Geophysical Research Letters*, **42**(5), 1606–1611.

Wang H., Fotidis I. A. and Angelidaki I. (2015). Ammonia effect on hydrogenotrophic methanogens and syntrophic acetate-oxidizing bacteria. *Fems Microbiology Ecology*, **91**(11), 1–8.

Weiland P. (2010). Biogas production: current state and perspectives. *Applied Microbiology and Biotechnology*, **85**(4), 849–860.

Westerholm M., Moestedt J. and Schnurer A. (2016). Biogas production through syntrophic acetate oxidation and deliberate operating strategies for improved digester performance. *Applied Energy*, **179**, 124–135.

Wunderlin P., Mohn J., Joss A., Emmenegger L. and Siegrist H. (2012). Mechanisms of N_2O production in biological wastewater treatment under nitrifying and denitrifying conditions. *Water Research*, **46**(4), 1027–1037.

Zeebe R. E. (2013). Time-dependent climate sensitivity and the legacy of anthropogenic greenhouse gas emissions. *Proceedings of the National Academy of Sciences of the United States of America*, **110**(34), 13739–13744.

Zumft W. G. (1997). Cell biology and molecular basis of denitrification. *Microbiology and Molecular Biology Reviews*, **61**(4), 533–616.

Chapter 2

Greenhouse gas emissions from wastewater treatment facilities

Xinmin Zhan

2.1 DIRECT AND INDIRECT GHG EMISSIONS

Direct emission of greenhouse gases (GHGs) in a wastewater treatment process unit is defined as the amount of these gases (CO_2, CH_4 and N_2O) directly emitted from the unit's water surface to the atmosphere. It is also known as fugitive emissions. Direct emission can be measured with certain techniques.

Indirect GHG emission refers to the carbon footprints associated with the operation of the wastewater treatment facilities due to the inputs of power and chemicals. Power is needed for running pumps and air blowers and for heating and lighting, and chemicals are needed for pH and alkalinity adjustment, phosphorus removal, coagulation/flocculation, and so on. Electricity and chemicals have intrinsic carbon footprints, which are the GHGs generated during their manufacturing and transport. The indirect emissions of GHGs are calculated using Eqs. 2.1 and 2.2:

$$CO_{2\text{-eq}} = E \times CO_{2\,\text{int},\,E} \tag{2.1}$$

$$CO_{2\text{-eq}} = \sum_{i} Q_i CO_{2\,\text{int},\,C,\,i} \tag{2.2}$$

where, $CO_{2\,\text{int},\,E}$ is the intrinsic equivalent CO_2 emission factor for electricity (g $CO_{2\text{-eq}}$/kWh); E is the amount of electricity consumed for running the facility (kWh); Q_i is the amount of chemical i used; and $CO_{2\,\text{int},\,C,i}$ is the intrinsic equivalent CO_2 emission factor for chemical i.

 Greenhouse Gas Emission and Mitigation in Municipal Wastewater Treatment Plants
Xinmin Zhan, Zhenhu Hu, Guangxue Wu
doi: 10.2166/9781780406312_017

The intrinsic carbon footprints of power and chemicals vary region by region, year by year, and wastewater treatment plants by plants. For municipal wastewater treatment plants (WWTPs), power is the major indirect GHG emission source. $CO_{2\ int,\ E}$ is dependent on the mixture of fuels that are used to produce electricity:

$$CO_{2\ int,\ E} = \sum F_i EF_i \tag{2.3}$$

where, F_i is the percentage contribution of fuel i in the fuel mixture to generate the electricity which is used by WWTPs; and EF_i is the CO_2 equivalent emission factor of fuel i when producing electricity (g $CO_{2\text{-}eq}$/kWh). F_i varies country by country, region by region and year by year; this leads to the variation of $CO_{2\ int,\ E}$. Table 2.1 lists EF_i values used in Ireland and Table 2.2 gives $CO_{2\ int,\ E}$ in Ireland in recent years. $CO_{2\ int,\ E}$ values are not constant due to the changing fuel mixture. Because of the use of renewable energy sources in producing electricity, $CO_{2\ int,\ E}$ values have significantly decreased recently.

Table 2.1 Carbon emission factors for fuels, EF_i, in Ireland (SEAI, Sustainable Energy Authority of Ireland, 2017).

Energy Type	**Kg $CO_{2\text{-}eq}$/kWh (2015)**
Net electricity imports	0.494
Offsite electric vehicle charging	0.529
Natural gas	0.205
LPG	0.229
Biogas, bioethanol and biodiesel	0.000
Kerosene	0.257
Gasoil	0.264
Light, medium & heavy fuel oils	0.274
Coal	0.341
Sod peat	0.374
Peat briquettes	0.356
Milled peat	0.414
Wood biomass (pellet, chips, logs, briquettes)	0.000
Solar thermal	0.000
Road diesel (DERV)	0.264
Petrol	0.252
Marked diesel (non-thermal use)	0.264
Jet A1 kerosene (aviation)	0.257
Aviation gasoline (AVGAS)	0.252

Table 2.2 Intrinsic CO_2 emission equivalent of electricity, $CO_{2\ int,\ E}$, in Ireland (SEAI, 2017).

Year	1999	2001	2003	2005	2007	2009	2011	2013	2015
kg $CO_{2\text{-eq}}$/ kWh	0.8149	0.8067	0.6746	0.6354	0.5602	0.5221	0.4889	0.4655	0.4675

A wastewater treatment plant in Dublin had an electricity bill of 14,820 MWh/yr in 2015. If the electricity was completely imported from the grid, then the indirect GHG emission due to the consumption of electricity in 2015 was equal to:

$$14{,}820\,\text{MWh/yr} \times 0.4675\,\text{kg}\,CO_{2\text{-eq}}/\text{kWh} \times 1000\frac{\text{kWh}}{\text{MWh}}$$
$$= 6{,}928{,}350\,\text{kg}\,CO_{2\text{-eq}}/\text{yr} = 6{,}928.4\,\text{tonne}\,CO_{2\text{-eq}}/\text{yr}$$

If the plant installed an on-site boiler burning woodchip biomass to produce 50% of the power required, then the indirect GHG emission was reduced to

$$14{,}820\,\text{MWh/yr} \times 50\% \times 0.4675\,\text{kg}\,CO_{2\text{-eq}}/\text{kWh} \times 1000\frac{\text{kWh}}{\text{MWh}}$$
$$= 3{,}464{,}175\,\text{kg}\,CO_{2\text{-eq}}/\text{yr} = 3{,}464{,}2\,\text{tonne}\,CO_{2\text{-eq}}/\text{yr}$$

This significant reduction would be due to 50% of the power used being generated from the renewable energy source, whose intrinsic carbon footprint is zero.

2.2 GHG EMISSION AND GENERATION

Greenhouse gases (CO_2, CH_4 and N_2O) are generated in wastewater treatment facilities due to biochemical reactions occurring, before they are emitted to the air from the aqueous phase. For a process unit element (Figure 2.1) for any gas *j*, a mass balance can be built in consideration of both the fluxes of gas generated in the unit element and the fluxes of gas emitted through the water-air interface.

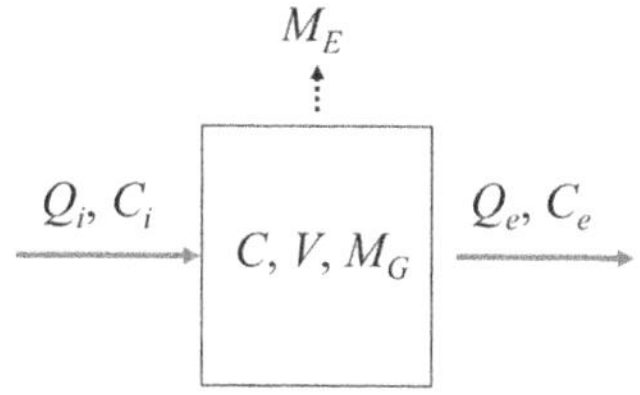

Figure 2.1 GHG mass balance in a process unit element.

$$\frac{dC}{dt}V = Q_iC_i - Q_eC_e + M_G - M_E \tag{2.4}$$

in which, Q_i and Q_e are the wastewater influent and effluent flow rates (m^3/s); C_i and C_e are the dissolved gas concentrations, for instance CH_4 gas, in the influent and effluent (kg/m^3); M_G is the flux of gas generated inside the unit element (kg/s); M_E is the flux of gas emitted from the unit element to the atmosphere from the element's water surface (kg/s); and V is the volume of the unit element.

At steady state, $Q_i = Q_e$, and $dC/dt = 0$, so

$$M_E = M_G + Q_iC_i - Q_iC_e \tag{2.5}$$

If $C_i \neq C_e$, then $M_E \neq M_G$, indicating that the fluxes of gas emission and the fluxes of gas generation are not necessarily equal. Therefore, when discussing the mechanisms of GHG generation in WWTPs, it is important to use GHG generation data rather than GHG emission data. Only using gas emission data would mislead the analysis.

If $C_i > C_e$, then $M_E > M_G$, indicating that GHG emission from this process unit is higher than *in situ* generation; If $C_i < C_e$, then $M_E < M_G$, indicating that GHG emission from this process unit is lower than *in situ* GHG generation. A municipal WWTP consists of a series of process units from pump stations to the discharge outlet. Some process units have $M_E < M_G$ and some have $M_E > M_G$ due to the operational conditions of the reactors which impact the biochemical reaction types and rates, and impact the mass transfer rates of the gases along the liquid-water interface.

2.3 EMISSION FACTORS OF GHG EMISSION

Equations 1.6 and 1.13 are the stoichiometric equations describing CH_4 and N_2O generation but in reality, when quantifying the emissions of CH_4 and N_2O, emission factors (EFs) are adopted in reference to the influent chemical oxygen demand (COD) or nitrogen (in most cases this is ammonium-nitrogen):

$$M_{CH_4} = EF_{CH_4} \times COD_i \times Q_i \tag{2.6}$$

$$M_{N_2O} = EF_{N_2O} \times N_i \times Q_i \times \frac{44}{28} \tag{2.7}$$

where M_{CH_4} and M_{N_2O} are the fluxes of emissions of CH_4 and N_2O from wastewater treatment facilities (kg CH_4/s and kg N_2O/s); EF_{CH_4} and EF_{N_2O} are the emission factors for CH_4 and N_2O, respectively (kg CH_4/kg COD_i and kg N_2O-N/kg N_i); COD_i is the influent COD concentration (kg COD/m^3 wastewater); and N_i is the influent nitrogen concentration (kg N/m^3 wastewater). The value 44/28 is used to convert mass of N (atomic weight of 14) to mass of N_2O (molecular weight of 44).

If the emission factor for N_2O is 11.5% for an intermittently aerated sequencing batch reactor treating a wastewater containing 2000 g N/m^3 of wastewater, then its N_2O emission from per m^3 wastewater treated is equal to:

$$2000\frac{\text{g}}{\text{m}^3} \times 11.5\% \times \frac{44}{28} = 361.4\frac{\text{g}}{\text{m}^3}$$

Taking the global warming potential (GWP) of N_2O into account, we can calculate CO_2 equivalent of N_2O emission, which is equal to:

$$361.4\frac{\text{g}}{\text{m}^3} \times 265 = 95{,}771\frac{\text{g}\,\text{CO}_{2\text{-eq}}}{\text{m}^3} = 95.8\frac{\text{kg}\,\text{CO}_{2\text{-eq}}}{\text{m}^3}$$

So, for treating 1 m^3 of this wastewater, 95.8 kg $CO_{2\text{-eq}}$ will be emitted from the treatment facility. This number is very alarming.

The emission factors can also be expressed as the amount of emissions relative to per capita, per m^3 wastewater treated, per tank surface area, etc. This makes the comparison of GHG emissions among different studies quite difficult.

2.4 EMISSIONS OF DISSOLVED GAS TO THE ATMOSPHERE

The principles explaining the emissions of dissolved CH_4 and N_2O gases into the atmosphere are (1) diffusion; and (2) air stripping. CH_4 has a poor solubility in water but N_2O is quite soluble in comparison with O_2. Table 2.3 compares the solubility of these two gases with O_2 and CO_2.

Table 2.3 Solubility of gases in water (pressure: 1 atm; unit: mg/L).

Temperature (°C)	5	10	15	20	25	30	35	40
CH_4	35.8	31.3	27.8	24.9	22.7	20.9	19.4	18.2
N_2O	2079	1727	1454	1239	1068	930	818	727
O_2 (pure)	61.5	54.6	49.0	44.5	40.8	37.7	35.2	33.2
O_2 (air)	11.3	10.1	9.1	8.3	7.6	7	6.5	6
CO_2			2007	1728	1503	1322	1173	

Source: Gevantman (2000).

The mass transfer of the gases from the liquid phase to the atmosphere across the water-air interface is postulated with a two-film theory. According to this theory, a thin liquid film exists in the liquid side and a thin gas film exists in the atmosphere side (Seader & Henley, 2006), as shown in Figure 2.2.

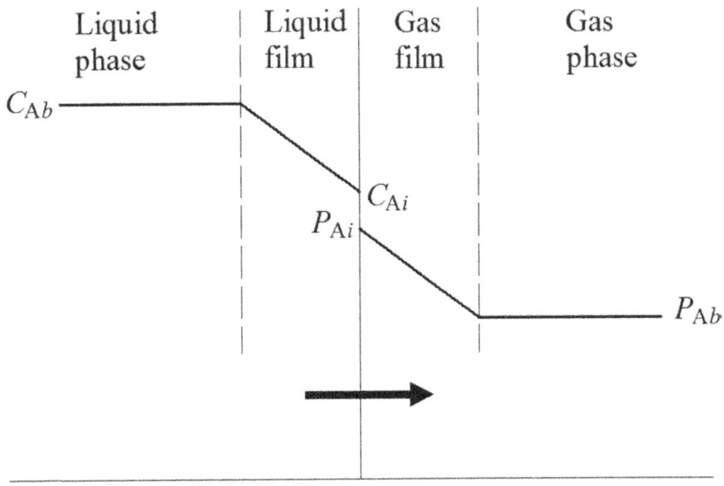

Figure 2.2 Two-film theory describing the mass transfer between the water and the atmosphere through the water-air interface (C_{Ab} and C_{Ai} are soluble gas concentrations in the bulk water phase and the gas-liquid interface, respectively; P_{Ab} and P_{Ai} are gas partial pressure in the bulk gas phase and the gas-liquid interface, respectively).

The mass transfer rates in the gas phase and the water phase at steady state, r, are written as:

In the gas phase, $r = K_{Ga} (P_{Ai} - P_{Ab})$ (2.8)

In the liquid phase, $r = K_{La} (C_{Ab} - C_{Ai})$ (2.9)

K_{La} and K_{Ga} are mass transfer coefficients in the liquid phase and the gas phase, respectively.

When the mass transfer in the gas phase and in the liquid phase reaches equilibrium then,

$$r = K_{Ga}(P_{Ai} - P_{Ab}) = K_{La}(C_{Ab} - C_{Ai}) \quad (2.10)$$

The dissolved gas concentration at equilibrium between the gas and water phases follows Henry's Law:

$$C_{Ai} = H \times P_{Ai} \quad (2.11)$$

If the wastewater treatment facilities are open to air, since the partial pressures of CH_4 and N_2O gases in the atmosphere are quite low, 0.00018% for CH_4 and 0.00003% for N_2O, the resistance in the gas phase is neglected in comparison with the resistance in the liquid phase for the two gases, then $P_{Ab} = P_{Ai} \approx 0$, so $C_{Ai} \approx 0$.

In this case, the expression of the mass transfer rates of CH_4 and N_2O from the wastewater to the air are simplified as:

$$r = K_{La}(C_{Ab} - C_{Ai}) = K_{La}C_{Ab} \tag{2.12}$$

K_{La} is also known as the overall mass transfer coefficient of CH_4 and N_2O from the water phase to the air. It varies depending on the wastewater characteristics, temperature at which the wastewater is treated, local turbulence and the air flow rate if aeration is provided (like in the aerated grit chamber and the aerated activated sludge tanks). The aeration provided into the water phase causes the air stripping effects.

K_{La} in wastewater can be correlated with K_{La} in clear water with a coefficient of α:

$$K_{La,\text{wastewater}} = \alpha K_{La,\text{clearwater}} \tag{2.13}$$

K_{La} values of volatile organic compounds (VOCs) along the water-gas interface in the wastewater are approximately the same as in the tap water test (Bielefeldt & Stensel, 1999), so α is thought to be 1 for VOCs. Therefore, it is quite reasonable to assume that for CH_4 and N_2O, a can also be assumed to be 1. This methodology has been used by several researchers in their research on studying N_2O emissions from wastewater (Li, 2010; Zhang *et al.* 2012; Pan *et al.* 2014).

Equation 2.5 can be converted to:

$$r_G V - K_{La}CV + QC_i - QC_e = 0 \tag{2.14}$$

r_G is the gas generation rate in the process unit element (kg/(m$^3 \cdot$ s)).

Hence,

$$C_e = \frac{QC_i + r_G V}{Q + K_{La}V} = \frac{C_i + \tau r_G}{1 + \tau K_{La}} \tag{2.15}$$

where, $\tau = V/Q$, the hydraulic retention time of the unit element. If K_{La} and the gas generation rate (r_G) are known, the concentration of the gas in the effluent from the unit element can be calculated.

2.5 GHG GENERATION AND EMISSION IN WWTPs

The development of wastewater treatment technologies and facilities keeps up with the development of civilization and our understanding of environment, ecology, and human health. To begin with, simple sedimentation was used to treat wastewater but today, complicated and expensive tertiary wastewater treatment processes are commonly required in most developed and developing countries. Figure 2.3 illustrates a typical flow chart of municipal WWTPs with tertiary treatment. Since CO_2 is not included in the GHG emission inventory in municipal WWTPs suggested by the Intergovernmental Panel on Climate Change (IPCC), only generation and emission of CH_4 and N_2O are briefed for each processing unit.

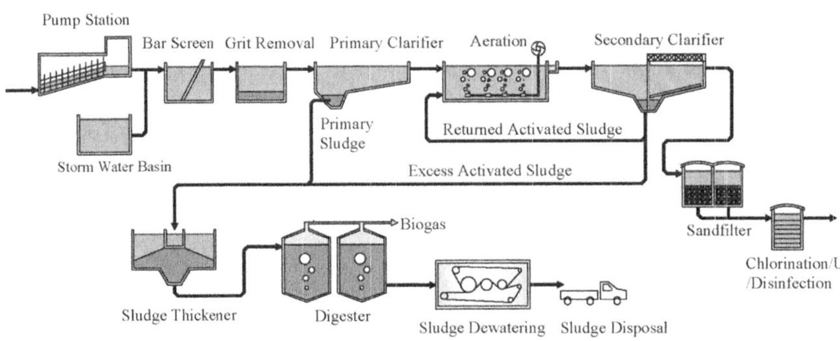

Figure 2.3 Flowchart of a typical municipal WWTP with tertiary treatment.

(1) *Pump station.* CH_4 is generated in the main sewer system, where anaerobic conditions are easily formed. Methane, up to 20–120 mg COD/L equivalent, can be formed in sewer systems (Guisasola *et al.* 2008; Guisasola *et al.* 2009). Therefore, a substantial emission of CH_4 might be observed in the pump station due to the turbulence.

(2) *Preliminary treatment.* Preliminary treatment process units are screening (a device with uniform-size openings to remove relatively large objects in the influent wastewater, like debris, plastic bags, etc.), and grit removal chambers which are designed to remove fast settleable inorganic particles, like sand and gravels. They are designed to have short hydraulic retention times (HRTs), for instance, grit chambers having HRTs of 30 seconds – 3 minutes (Tchobanoglous *et al.* 2003). Hence, the generation of CH_4 and N_2O in these two units is negligible. However, like in the pump station, CH_4 generated in the main sewer system would be emitted in the preliminary treatment process units due to turbulence. If grit chambers are aerated, where air is blown to prevent

the settling of light organic particles, CH_4 emission would be quite high due to the air stripping effect.

(3) *Primary clarification.* Primary clarifiers are designed to remove inorganic and organic suspended solids from wastewater. In the primary clarifiers, due to the relatively long retention time, up to 2 hours typically, a certain degree of anaerobic condition would be developed, causing the generation of CH_4. N_2O generation would be low due to the low dissolved oxygen and biomass concentrations in the wastewater, limiting the occurrence of nitrification. The gases generated in primary clarifiers would be emitted in the following aeration tanks due to air stripping.

(4) *Secondary treatment.* In secondary treatment, microorganisms are cultured by a variety of engineering methods for decomposition of organic matter (represented by COD, and 5-day biochemical oxygen demand (BOD_5)) in aerobic or anaerobic conditions, oxidization of NH_4^+ in aerobic conditions (i.e., nitrification), reduction of oxidized nitrogen (NO_2^- and NO_3^-) to N_2 gas in anoxic conditions (i.e., denitrification), and luxury accumulation of phosphorus (i.e., enhanced biological phosphorus removal) in alternative anaerobic/aerobic conditions. These biochemical reactions generate CH_4 and/or N_2O. The emission of these two gases in WWTPs mainly takes place in the secondary treatment. An example of secondary wastewater treatment is the conventional A^2O (anaerobic/anoxic/oxic) process (Figure 2.4), which consists of the anaerobic, anoxic and aerobic conditions for achieving COD, nitrogen and phosphorus removal.

In the anaerobic tank, because of the presence of anaerobic conditions, organic matter is probably reduced to CH_4, and NO_3^- which is brought in with the return sludge is probably reduced to N_2O. In the anoxic tank, N_2O generation would be more substantial than in the anaerobic tank because of the dominating denitrification reaction. The low dissolved oxygen (DO) concentration would also benefit the generation of CH_4.

In the aerobic tank, N_2O is generated due to nitrifier denitrification and hydroxylamine oxidation, while CH_4 generation is not significant under the aerobic conditions. Meanwhile, the air stripping effect enhances the emissions of N_2O and CH_4, which are generated in the proceeding process units and in-lane of the aerobic tanks to the air. In the secondary clarifier, a relatively long retention time would cause CH_4 and N_2O to form, in particular if the secondary clarifier is wrongly designed and the sludge biomass remains in the clarifier for too long. One indication of the wrong design is the floating biomass caused by N_2 gas or CH_4 gas bubbles produced in the clarifier.

(5) *Tertiary treatment.* Tertiary treatment is increasingly adopted in WWTPs for further reduction of N and P from wastewater. If chemicals (like aluminum and iron salts) are added to precipitate phosphorus, the amount of N_2O and CH_4 generation is quite small in the chemical process units.

However, if post-denitrification activated sludge or biofilters are used, in which external carbon, like methanol, is added to supply organic matter to denitrifiers to denitrify NO_3^- in the wastewater, then N_2O production is inevitable. In addition, if the external organic carbon source is from the fossil origin, conversion of this organic matter to CO_2 should be included in the GHG emission inventory of the WWTPs.

(6) *Sludge treatment.* After thickening, the sludge (primary and secondary sludge) is dewatered followed by thermal treatment (gasification, pyrolysis, combustion), land-spreading, composting, or anaerobic digestion. The dewatered liquor is recirculated and treated in the mainstream wastewater treatment process. Because the oxidized nitrogen concentrations are low, N_2O emission is small during sludge thickening and dewatering. However, due to the niche of anaerobic conditions, significant CH_4 may be generated. Anaerobic digestion of sludge, producing biogas from the organic matter in sludge, provides a renewable energy source, which is commonly used to offset the energy use imported from the grid, but a certain amount of CH_4 is lost from the anaerobic digesters (due to gas leaking), which also contributes to the GHG emissions of municipal WWTPs. GHG emissions (both CH_4 and N_2O) from land-spreading and thermal treatment of sludge are outside the content of this book, but the methods to calculate GHG emissions of sludge land-spreading and thermal treatment, suggested by the IPCC, will be described in Chapter 3.

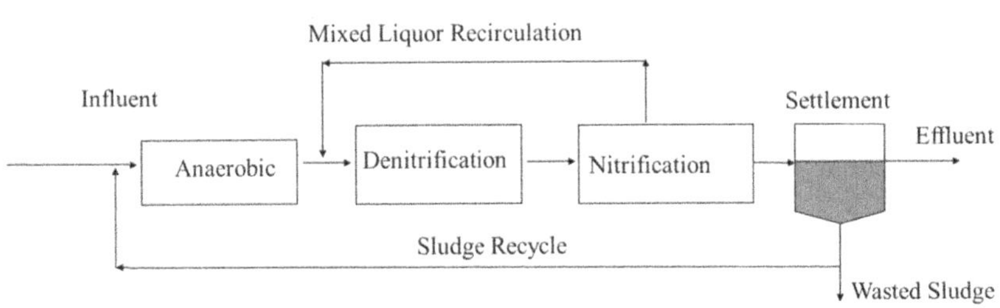

Figure 2.4 Schematic of A^2O process.

2.6 SUMMARY

In municipal WWTPs, due to the complexity of biochemical reactions, CH_4 and N_2O are often formed in individual wastewater treatment process units. We must bear in mind when discussing GHG emissions from WWTPs that emission and generation are not synchronic. The generation, emission and possible mitigation of CH_4 and N_2O from wastewater treatment plants are summarized in Table 2.4.

Table 2.4 Generation and emission of CH_4 and N_2O from wastewater treatment facilities.

Process Unit	CH_4		N_2O		Reason	Mitigation
	Generation	Emission	Generation	Emission		
Pump station	N	Yes	N	N	Flow induced turbulence; air stripping effects	
Screen	N	Yes	N	N	Turbulence	
Grit chamber	N	Yes	N	N	Flow induced turbulence; air stripping effects in aerated grit chambers	
Primary clarifier	N	Yes	N	N	Diffusion; flow induced turbulence	
Secondary wastewater treatment tank	Yes	Yes	Yes	Yes	Generation due to biochemical reactions; diffusion; local turbulence; air stripping in aeration tanks and biofilters	Use proper process design and operation procedure
Secondary clarifier	Yes	Yes	Yes	Yes	Generation due to biochemical reactions; diffusion; local turbulence	Proper process design and operation
Sludge thickening and holding tank	Yes	Yes	N	N	Generation due to biochemical reactions; diffusion; local turbulence	Proper process design and operation
Sludge anaerobic digesters	Yes	Yes	N	N	Methane generation and loss from digesters	Reduce methane loss

Note: N, negligible.

2.7 REFERENCES

Bielefeldt A. R. and Stensel H. D. (1999). Treating VOC-contaminated gases in activated sludge: mechanistic model to evaluate design and performance. *Advances in Environmental Science and Technology*, **33**(18), 3234–3240.

Gevantman L. (2000). Solubility of selected gases in water. *Nitric Oxide (NO)*, **308**(3.348), 10–14.

Guisasola A., de Haas D., Keller J. and Yuan Z. (2008). Methane formation in sewer systems. *Water Research*, **42**(6), 1421–1430.

Guisasola A., Sharma K. R., Keller J. and Yuan Z. (2009). Development of a model for assessing methane formation in rising main sewers. *Water Research*, **43**(11), 2874–2884.

Li J. (2010). Nutrient Removal from Slaughterhouse Wastewater and Oxygen-Limited Partial Nitrification Process in Sequencing Batch Reactors. National University of Ireland, Galway.

Pan M., Wen X., Wu G., Zhang M. and Zhan X. (2014). Characteristics of nitrous oxide (N_2O) emission from intermittently aerated sequencing batch reactors (IASBRs) treating slaughterhouse wastewater at low temperature. *Biochemical Engineering Journal*, **86**, 62–68.

Seader J. D. and Henley E. J. (2006). Separation Process Principles, 2nd edn. Wiley, Hoboken, NJ.

SEAI (2017). Emission Factors. Sustainable Energy Authority of Ireland.

Tchobanoglous G., Burton F. L., Stensel H. D. and Metcalf and Eddy Inc. (2003). Wastewater Engineering: Treatment and Reuse. McGraw-Hill Education, New York.

Zhang M., Lawlor P. G., Li J. and Zhan X. (2012). Characteristics of nitrous oxide (N_2O) emissions from intermittently-aerated sequencing batch reactors treating the separated liquid fraction of anaerobically digested pig manure. *Water, Air & Soil Pollution*, **223**(5), 1973–1981.

Chapter 3

IPCC framework for calculating greenhouse gas emissions from wastewater treatment

Conor Dennehy and Xinmin Zhan

3.1 INTRODUCTION

Concerns in relation to anthropogenic greenhouse gas (GHG) emissions have resulted in major investment into the development of renewable energy technologies, in thorough studies of carbon flows and in development of policy instruments which incentivize a reduction in GHG emissions. Due to its ubiquitous nature, wastewater treatment has the potential to be a significant source of GHG emissions globally.

As discussed in the previous two chapters, the handling, transport, treatment and discharge of wastewater leads to the release of CH_4 and N_2O gases. Quantifying these emissions is essential in order to determine the effect of wastewater treatment infrastructure on national GHG inventories. The Intergovernmental Panel on Climate Change (IPCC) provides guidelines for the estimation of national GHG emissions relevant to GHG emissions from wastewater treatment.

The IPCC is a United Nations (UN) mandated international organization tasked with providing reports on the effects of anthropogenic GHG emissions and providing guidelines for estimating national GHG emissions. While not a research body itself, the IPCC collates published work in the area of climate change in order to provide reports designed to support policymakers operating in the area of climate change. The IPCC Guidelines for National Greenhouse Gas Inventories is broken into five volumes (IPCC, 2006):

(1) General Guidance and Reporting
(2) Energy
(3) Industrial Processes and Product Use
(4) Agriculture, Forestry and Other Land Use
(5) Waste

Xinmin Zhan, Zhenhu Hu, Guangxue Wu
doi: 10.2166/9781780406312_029

Each of these volumes provides a detailed description of methods by which the annual national GHG emissions from various activities can be calculated. Guidelines on how to estimate GHG emissions from wastewater treatment are contained in Volume 5; however the volume devoted to emissions from wastewater treatment does not account for all aspects of the GHG emission potential of wastewater treatment. For example, the GHG emissions associated with utilizing CH_4 produced from anaerobic digestion of waste sludge generated in wastewater treatment plants (WWTPs) is calculated using formulae found in Volume 2 (Energy), while the GHG emissions from land application of sludge (a type of biosolids) is estimated using formulae found in Volume 4. While such subdivision makes sense when calculating national GHG inventory, it makes estimating GHG emissions from the entire municipal WWTPs somewhat complex. The following sections bring together all the relevant IPCC guidelines pertaining to GHG emissions from wastewater treatment processes.

3.2 GREENHOUSE GAS EMISSIONS FROM WASTEWATER TREATMENT AND DISCHARGE

The IPCC guidelines not only account for GHG emissions from wastewater treatment, but also account for emissions from wastewater (treated and untreated) after discharging into the environment. Figure 3.1 illustrates the types of wastewater activities accounted for by the IPCC guidelines.

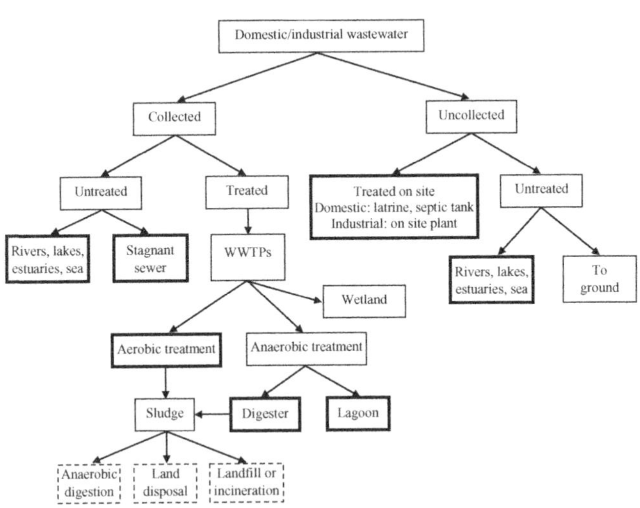

Figure 3.1 Wastewater discharge and treatment pathways identified by the IPCC guidelines (IPCC, 2006). Items in black bold boxes have emissions accounted for by IPCC guidelines on emissions from wastewater treatment. Items in dashed boxes have emissions accounted for and reported in chapters other than the Wastewater Treatment chapter. *Note*: WWTPs indicates wastewater treatment plants.

The IPCC guidelines do not account for CO_2 emissions from wastewater treatment and discharge. As such emissions are deemed to be of biogenic origin, they are not included in national inventories. Therefore, the greenhouse gases of concern in relation to wastewater treatment and discharge are CH_4 and N_2O. The IPCC provides two separate frameworks for determining CH_4 and N_2O emissions, and separate methodologies for calculating CH_4 emissions from domestic and industrial wastewater treatment and discharge.

3.3 IPCC GUIDELINES FOR THE ESTIMATION OF NATIONAL GHG EMISSIONS OF WASTEWATER TREATMENT

The IPCC specifies a framework for calculating GHG emissions from wastewater treatment in Volume 5, Chapter 6 of the 2006 guidelines. The framework provides three separate calculation methodologies (referred to as 'Tiers'). The choice of methodology is decided by the level of information available in the country in which GHG emissions are to be calculated. Figure 3.2 presents the decision tree provided by the IPCC for determining which Tier should be used when estimating GHG emissions from wastewater treatment, based on the information available.

As can be seen from Figure 3.2, in order to utilize the IPCC methodology to estimate national GHG emissions from wastewater treatment, it is essential to obtain data relating to the types and prevalence of various wastewater treatment processes applied. Without this information, no estimates can be made. If there is not a sufficient amount of data to merit a bottom-up approach, or if no specific emission factors have been calculated for the country in question, then the Tier 1 approach can be taken (provided that emissions from wastewater treatment are deemed not to have a significant effect on the country's overall GHG inventory). This approach combines the country-specific data relating to the types of wastewater infrastructure used, with default emission factors provided in the IPCC methodology. This approach can allow for GHG emission estimates to be made without significant primary research being completed on the emission factors observed in specific countries. However, because of this, there is a reliance on the default emission factors provided, which may be less accurate than the emission factors determined by local field research. It is therefore only regarded as good practice to use such an approach in countries where limited data is available. The Tier 1 methodology does not account for the effects of sludge disposal or the potential mitigation effects of energy recovery through biogas utilization.

The Tier 2 approach is undertaken if country emission factors are available for specific wastewater treatment processes. Such country-specific emission factors are typically determined through field-scale monitoring. In addition, the Tier 2 approach can also account for GHG emissions from sludge disposal and biogas utilization, provided the additional data (sludge disposal routes and rate of biogas utilization) is available. However, in terms of national GHG inventories, these emissions should be reported in separate categories (i.e., biogas utilization in Energy, and emissions from land application in Land Use). The Tier 3 approach is used when a large amount of granular data in relation to the GHG emission rates based on regional and/or plant specific data are available. Due to the large amounts of data required, few countries have attempted such a bottom-up approach to assessing GHG emissions from wastewater treatment.

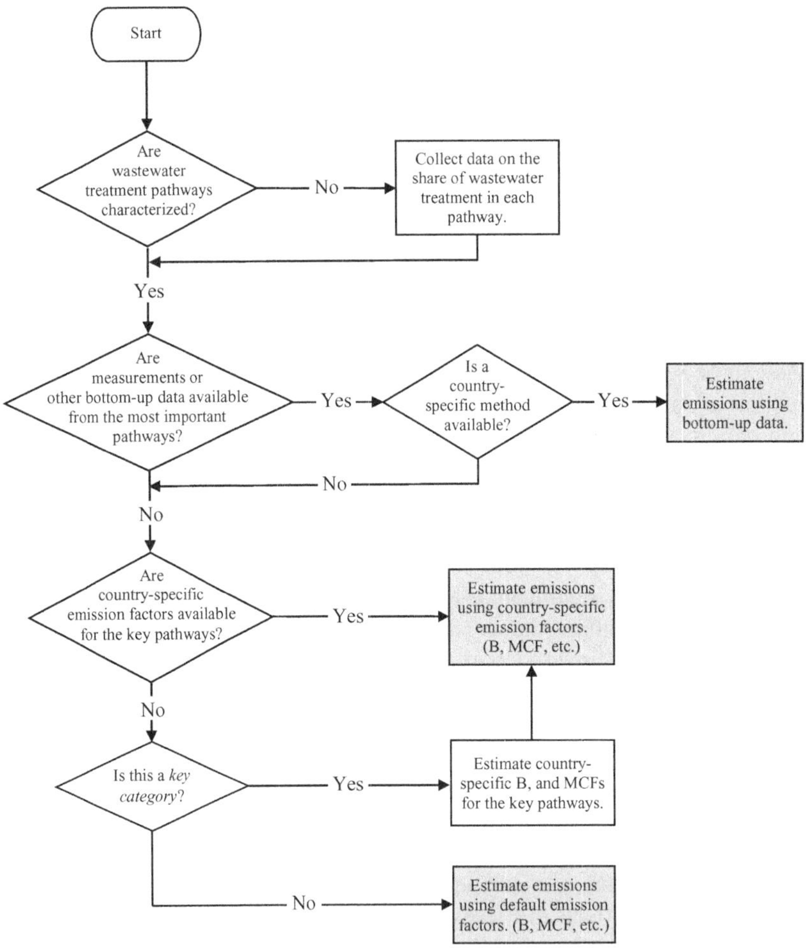

Figure 3.2 Decision tree for determining the approach when calculating GHG emissions based on IPCC methodology (IPCC, 2006). A key category is one that has significant contribution to a country's national GHG inventory and should be prioritized within the national inventory system (IPCC, 2006). Note that B and MCF, emission factors, are defined in the text in 3.3.1.

The equations used to estimate GHG emissions using Tier 1 and Tier 2 methodology are generally comprised of two factors:

(1) An activity factor which is the rate at which an activity occurs, e.g., amount and concentration of wastewater treated, population having centralized wastewater treatment etc.
(2) An emission factor which determines the rate of emission of a GHG based on the activity undertaken, e.g., kg $CO_{2\text{-eq}}$ per m^3 wastewater treated.

Each of these factors varies for each stage of the wastewater treatment process, where different equations are used to determine the overall GHG emissions from wastewater treatment. The IPCC provides default emission factors in most cases; however, it suggests the use of country specific factors when available.

In order to fully assess the GHG emissions from wastewater treatment, the GHG emissions resulting from various activities (as defined by the IPCC) must be calculated. Figure 3.3 illustrates the IPCC defined GHG generating activities during wastewater treatment. How the annual GHG emissions from each of these activities are calculated will be outlined in the following sections.

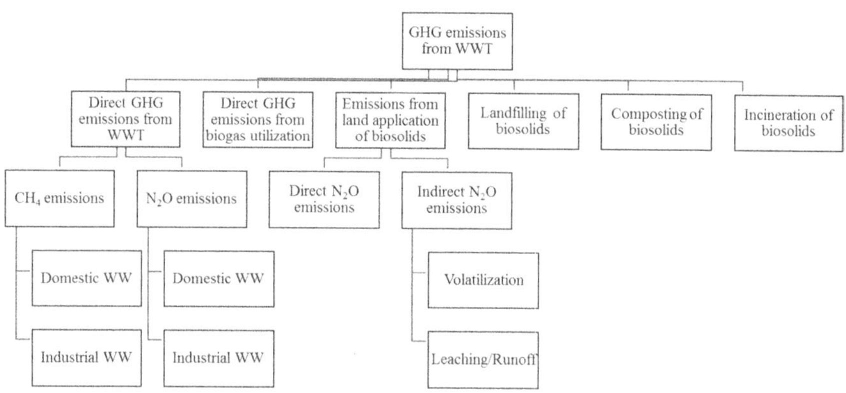

Figure 3.3 Illustration of the GHG emitting wastewater treatment related activities. The GHG emissions from each of these activities should be quantified to estimate the GHG emissions from wastewater treatment.

3.3.1 CH_4 emissions from domestic wastewater treatment

The equation provided by the IPCC (IPCC, 2006) to estimate national CH_4 emissions from domestic wastewater treatment and discharge is provided in Eq. 3.1.

$$CH_{4,\text{Emission}} = \left[\sum_{i,j} (U_i \cdot T_{i,j} \cdot EF_j)\right](TOW - S) - R \tag{3.1}$$

where

$CH_{4,Emission}$ = CH_4 emissions in inventory year (kg CH_4/yr);
TOW = total organics in wastewater in inventory year (kg BOD/yr);
S = organic component removed as sludge in inventory year (kg BOD/yr);
U_i = fraction of population in income group i in inventory year;
$T_{i,j}$ = degree of utilization of treatment/discharge pathway or system, j, for each income group fraction i in inventory year;
i = income group: rural, urban high income and urban low income;
j = each treatment/discharge pathway or system;
EF_j = emission factor for treatment system j (kg CH_4/kg BOD); and
R = amount of CH_4 recovered in inventory year (kg CH_4/yr).

Since CH_4 emissions are due to the conversion of biologically degradable carbon to CH_4 via methanogenic organisms, the concentration of total organic carbon in wastewater (TOW) is critical in determining the GHG emissions. Therefore, TOW determination is the first step in estimating the national CH_4 emissions from wastewater treatment and discharge. TOW is a function of population. The IPCC provides an equation for the calculation of TOW based on population, per capita biochemical oxygen demand (BOD) loading and a factor accounting for industrial discharges to municipal WWTPs. It is recommended that a country specific per capita BOD loading rate is utilized, however the IPCC provides default values for a range of countries and regions. The factor S in Eq. 3.1 can be determined through estimating the level of BOD removal and sludge production in each treatment system. The R value accounts for CH_4 captured during the wastewater treatment process. This factor is the means by which the guidelines account for anaerobic systems where CH_4 is flared or utilized as an energy source (as opposed to resulting in direct CH_4 emissions).

In addition to data on the types and prevalence of various wastewater management infrastructures, the guidelines require data on the fraction of the population in specific income groups. The three groups specified are rural, urban high income and urban low income. The rationale for requiring this distinction is that wastewater treatment methods vary significantly depending on income background and where people live. Therefore, utilizing factors accounting for this distinction provides greater accuracy in the final calculation. The IPCC provides suggested values for U_i and T_{ij} for a range of countries, compiled through the use of population databases (for U_i) and expert opinion (T_{ij}).

The IPCC provides an equation for the calculation of the emission factor EF_j (Eq. 3.2) for domestic wastewater. The emission factor is determined by the inherent maximum CH_4 producing capacity of wastewater (B_o) and the level at which this CH_4 is emitted (which is dependent on treatment system type).

$$EF_j = B_o \times MCF_j \tag{3.2}$$

where

B_o = maximum CH_4 producing capacity (kg CH_4/kg BOD); and

MCF_j = methane correction factor.

From Eq. 1.13, B_o is calculated to be 0.6 kg CH_4/kg BOD, which is provided by the IPCC as a default value. The guidelines suggest that the best practice is to utilize experimentally derived national B_o values. The IPCC provides MCF values for arrange of wastewater treatment systems (Table 3.1).

Table 3.1 CH_4 correction factors provided by IPCC guidelines.

Type of Treatment Systems and Discharge Pathways	Comments	MCF	Range
Untreated systems			
Sea, river and lake discharge	Anaerobic conditions can exist with high organic loadings.	0.1	0–0.2
Stagnant sewer	Open and warm.	0.5	0.4–0.8
Fast flowing sewer (open or closed)	Insignificant CH_4 emission from pump stations, etc.	0	0
Treated systems			
Well managed centralized, aerobic system	CH_4 can be emitted from clarifiers and other zones where anaerobic conditions are present.	0	0–0.1
Poorly centralized, overloaded aerobic system	Overloaded.	0.3	0.2–0.4
Anaerobic digesters for sludge management	If no CH_4 recovery.	0.8	0.8–1.0
Anaerobic treatment system	If no CH_4 recovery.	0.8	0.8–1.0
Anaerobic shallow lagoon (depth < 2 metres)	Expert judgement is required.	0.2	0–0.3
Anaerobic deep lagoon (depth > 2 metres)		0.8	0.8–1.0
Septic tank	Half of BOD_5 settles in anaerobic conditions.	0.5	0.5
Latrine in dry climate, and for a small family (3–5 persons)	Groundwater table lower than latrine.	0.1	0.05–0.15
Latrine in dry climate, and for communal use (many users)	Groundwater table lower than latrine.	0.5	0.4–0.6
Latrine in wet climate	Use flush water and, groundwater table higher than latrine.	0.7	0.7–1.0
Latrine with regular removal of sediment	Sediment removal for fertilizer.	0.1	0.1

Source: IPCC (2006).

It is clear that the highest MCFs are attributed to anaerobic processes, while aerobic systems only release CH_4 if poorly managed. The high MCFs provided for the use of anaerobic processes may not in many cases reflect the true GHG emission potential as the capture and flaring or utilization of CH_4 is not included in the emission factor. The IPCC guidelines provide the *R* factor in Eq. 3.1 to account for instances where the CH_4 generated by anaerobic processes is captured and utilized or flared. This approach is used to account for fugitive emissions from anaerobic processes, i.e., where CH_4 emissions generated from the anaerobic process (based on emission factors) are greater than the amount of CH_4 captured or flared.

Septic tank systems can be a significant source of CH_4, with an MCF greater than even poorly managed centralized WWTPs. Latrines can be a significant source of CH_4 depending on the size and level of maintenance. The MCFs presented in Table 3.1 relating to untreated wastewater discharge, and relating to latrine and septic tanks indicate that countries which have a decentralized or poorly developed wastewater treatment infrastructure may face greater GHG emissions than countries where modern centralized WWTPs using aerobic wastewater treatment processes are built (however this is dependent on whether the aerobic systems are well managed, or whether the biogas produced from anaerobic digestion of sludge is utilized).

3.3.2 CH_4 emissions from industrial wastewater treatment

Similar to domestic wastewater, CH_4 emissions will occur if industries treat their wastewater with a high organic loading rate. The IPCC suggests, provided that the granular emission data are available, a bottom-up (Tier 3) approach should be used for determining CH_4 emission from industrial wastewater treatment. In lieu of this, the guidelines provide an equation (Eq. 3.3) to calculate CH_4 emissions from industrial wastewater treatment which is similar to the equation for domestic wastewater (Eq. 3.1). However, due to the variations in *TOW, S, EF* and *R* within the industrial wastewater sector, the IPCC guidelines suggest provision of specific *TOW* and *S* values for each industry. These data might be easier to obtain than those for domestic wastewater, since in most countries large industries are required to report the wastewater influent and emission data and wastewater treatment facilities' operational data to the national environmental protection agencies.

$$CH_{4,\text{Emission}} = \sum_i [(TOW_i - S_i) \times EF_i - R_i] \quad (3.3)$$

where

TOW_i = total organically degradable material in wastewater from industry *i* in inventory year (kg COD/yr);

i = industry sector;

S_i = organic component removed as sludge in inventory year (kg COD/yr);

EF_i = emission factor for industry i (kg CH_4/kg COD). If more than one treatment practice is used in an industry this factor would need to be a weighted average; and

R_i = amount of CH_4 recovered in inventory year (kg CH_4/yr).

It should be noted that, unlike domestic wastewater, the loading rates and emission factors are calculated using chemical oxygen demand (COD) rather than BOD because COD is widely used in industries.

3.3.3 N_2O emissions

Unlike the guidelines provided for estimating CH_4 emissions from wastewater treatment, the guidelines for estimating N_2O emissions do not provide a tiered approach because of limited research on N_2O emission from WWTPs. While the IPCC does provide guidelines on direct N_2O emission from wastewater treatment (Eq. 3.4), they regard direct emissions as a minor N_2O source, therefore suggesting that such emissions need only be estimated in countries which manage wastewater primarily through centralized WWTPs. The IPCC does not provide guidance for the estimation of direct N_2O emissions from decentralized wastewater treatment systems (septic tanks and latrines) possibly because of limited nitrification and denitrification in these two systems. The guidelines suggest using Eq. 3.4 as the best practice when estimating national N_2O emissions. Equation 3.4 is an estimate of emissions from treated wastewater discharged to the environment rather than emissions associated with the treatment of wastewater.

$$N_2O_{\text{Emission}} = N_{\text{Effluent}} \times EF_{\text{Effluent}} \times \frac{44}{28} \tag{3.4}$$

where

N_2O_{Emission} = N_2O emissions in inventory year (kg N_2O/yr);

N_{Effluent} = nitrogen in the effluent discharged to aquatic environments (kg N/yr);

EF_{Effluent} = emission factor for N_2O emissions from discharged wastewater (kg N_2O-N/kg N); and

44/28 is used to convert the mass of N to N_2O (atomic mass of N is 14 and relative molecular mass of N_2O is 44).

The N_{Effluent} value is calculated based on the population, the per capita discharge of protein entering the WWTPs and the amount of N removed from the wastewater in sludge:

$$N_{\text{Effluent}} = PE \times Ptein \times F_{\text{NPR}} \times F_{\text{Non-Con}} \times F_{\text{Ind-Com}} - N_{\text{Sludge}} \tag{3.5}$$

where

PE = population;

$Ptein$ = annual per capital protein consumption (kg/person/yr);

F_{NPR} = fraction of nitrogen in protein;
$F_{Non\text{-}Con}$ = factor for non-consumed protein added to the wastewater;
$F_{Ind\text{-}Com}$ = factor for industrial and commercial co-discharged protein into the sewer system; and
N_{Sludge} = nitrogen removed with sludge (kg N/yr).

The IPCC default value for F_{NPR} is 0.16 kg N/kg protein, and the default value for $N_{Sludge} = 0$. The IPCC guidelines provide a single default value for the $EF_{Effluent}$ factor of 0.005 kg N_2O-N/kg N. This value is based on a small amount of field observations and assumptions on N_2O emission from wastewater discharge to estuaries and rivers. The IPCC recognizes that because of the lack of field data, $EF_{Effluent}$ has a high degree of associated uncertainty.

In most developed countries, centralized WWTPs are the most prevalent means of managing wastewater. In these instances, the IPCC suggests that direct N_2O emission may be significant and provides a means of calculating direct N_2O emissions from wastewater treatment facilities. The equation (Eq. 3.6) provided for estimating direct N_2O emissions accounts for the population which utilizes centralized WWTPs, the nitrogen load from industrial and commercial wastewaters and the rate at which direct N_2O emissions occur in WWTPs.

$$N_2O_{Emission} = PE \times T_{Plant} \times F_{Ind\text{-}Com} \times EF_{Plant} \tag{3.6}$$

where
T_{Plant} = degree of utilization of modern, centralized wastewater treatment plants (%); and
EF_{Plant} = emission factor, 3.2 g N_2O/person/yr.

The IPCC default value for $F_{Ind\text{-}Com}$ is 1.25. The default emission factor provided was taken from the study of Czepiel *et al.* (1995) who completed field measurements of N_2O emissions from a US municipal WWTP. A high degree of uncertainty is associated with this emission factor due to reliance on a single set of observations.

If direct N_2O emissions from wastewater treatment are to be included in a national GHG inventory, then the amount of nitrogen in the estimated emissions must be subtracted from the $N_{Effluent}$ value which is used to determine national N_2O emissions in Eq. 3.4.

3.4 GREENHOUSE GAS EMISSIONS FROM BIOGAS UTILIZATION

The IPCC guidelines require the effect of biogas utilization to be accounted for in national GHG emissions from energy generation, rather than from emissions from wastewater treatment. This can result in the net effect of anaerobic processes within WWTPs not being fully accounted for outside of national GHG emission inventories. As biogas generated from wastewater treatment has a low carbon footprint when used as a fuel, it has the potential to offset the use of more carbon intensive fuels such as coal and oil. This in turn results in a net decrease in GHG emissions from energy generation at a national level. The IPCC accounts for this benefit by providing a lower default emission factor for biogas generated from the anaerobic digestion of wastewater treatment sludge than coal and oil (see Table 3.2).

Table 3.2 IPCC default emission factors for the use of biogas (from sludge anaerobic digestion), coal (anthracite) and crude oil in the energy industry.

Fuel Source	Energy Industry CO_2 Emission Factor (kg/TJ)	Energy Industry CH_4 Emission Factor (kg/TJ)	Energy Industry N_2O Emission Factor (kg/TJ)
Sludge gas	54,600	1	0.1
Anthracite	98,300	1	1.5
Crude oil	73,300	3	0.6

Source: IPCC (2006).

The IPCC provides three tiers to calculate GHG emissions from energy generation. Tiers 1 and 2 require the relative contribution of the fuel types used in the country for energy generation, and the CO_2, CH_4 and N_2O emission factors for each fuel (Eq. 3.7). The IPCC provides default emission factors, but recommends the use of country-specific emission factors if available. Tier 3 guidelines require a breakdown of the energy generating technologies used, the fuel type used and country-specific CO_2, CH_4 and N_2O emission factors.

$$\text{Emissions}_{\text{GHG}} = \text{Fuel consumption} \times EF \tag{3.7}$$

where

$\text{Emissions}_{\text{GHG}}$ = emissions of a given gas (kg);
Fuel consumption = the amount of fuel combusted (TJ); and
EF = default emission factor of a given gas by the fuel (kg gas/TJ).

When attempting to undertake an estimation of GHG emissions from wastewater treatment, either at plant scale or at a national level, using the methodology provided by the IPCC for assessing the effect of biogas utilization, will result in an increase in the estimated GHG emission from wastewater treatment. An adapted approach will be required if the mitigation effects of utilizing biogas as a fuel source are to be quantified. Such an approach will require assuming the displacement of fossil fuels currently used in the national fuel mix, with the emissions avoided dependent on the difference in emission factors between the biogas and the displaced fuels (for WWTPs, CO_2 emission factor is zero, since the biogas is biogenic).

It should also be noted that an alternate set of existing guidelines govern the GHG emissions from the anaerobic digestion of wastewater treatment sludge when it is removed from WWTPs and co-digested with other forms of waste as a means of energy generation. These guidelines are found in Volume 5, Chapter 4 of the IPCC guidelines.

3.5 GREENHOUSE GAS EMISSIONS FROM SLUDGE DISPOSAL

The IPCC guidelines do not include GHG emissions from the handling and disposal of WWTP sludge in the method for estimating GHG emissions from wastewater. Instead, GHG emissions from sludge disposal are collated depending on disposal routes. This has a significant effect on the calculated national GHG emissions from wastewater treatment, as GHG emissions from sludge handling and disposal can be as much as two orders of magnitude greater than the GHG emissions from wastewater treatment (Hobson, 2000). While the distinction between wastewater treatment and sludge disposal is logical in terms of estimation of national GHG inventories, it poses challenges when an estimation of the total GHG emissions from WWTPs is desired. The following sections detail the IPCC guidelines for calculating GHG emissions from sludge disposal. It should be noted that despite the fact that previous IPCC reports identify sludge storage as a potential source of GHG emissions, the IPCC does not directly provide guidelines for the calculation of GHG emissions for sludge disposal.

3.5.1 Land application

The application of wastewater treatment sludge to soil can result in GHG emissions. N_2O is the primary GHG of concern for land application of WWTP sludge. It is emitted directly due to the nitrification and denitrification processes in the soil after the application of sludge. It can also be emitted indirectly through the volatilization of NH_3 and NO_x gases, and subsequent deposition onto land and surface waters, and through leaching of the N content of sludge into water bodies (where subsequent transformations result in N_2O emission). While CH_4 emissions may occur if anaerobic conditions occur in the soil to which the sludge is applied, CH_4 emissions are not considered by the IPCC guidelines. The IPCC guidelines for estimating GHG emissions from the land application of sludge are found in Volume 4 of the guidelines (Agriculture, Forestry and Other Land Use). For the purpose of the methodology found in the IPCC guidelines, the WWTP sludge is classified as an organic soil amendment. Both direct and indirect N_2O emissions from organic soil amendments are estimated.

Three tiers of methodology are provided for the estimation of direct and indirect N_2O emissions from soil amendments. Tier 2 methodology is recommended when country-specific emission factors for various soils, and organic amendments are available. Tier 3 methodology is recommended only when experimentally validated models for N_2O emissions in the target country and country-specific emission factors are available. The Tier 1 equation uses default emission factors to estimate N_2O emissions from managed soils. In terms of assessing N_2O emissions from land application of sewage sludge, these can be calculated using Eq. 3.8.

$$N_2O_{\text{Direct sludge}} = F_{\text{SEW}} \times EF \times \frac{44}{28} \tag{3.8}$$

where

$N_2O_{\text{Direct sludge}}$ = annual direct N_2O emissions due to land application of sewage sludge in inventory year (kg N_2O/yr);

F_{SEW} = annual amount of nitrogen in sewage sludge applied to land (kg N/yr); and

EF = emission factor for N_2O emissions from N inputs (kg N_2O–N/kg N).

The IPCC provides a default emission factor of 0.01 kg N_2O–N/kg N for the calculation of N_2O emissions from organic fertilizers, however countries are encouraged to determine country-specific emission factors. The IPCC provides two Tier 1 equations for the calculation of indirect emissions; one for atmospheric indirect emissions, and the other one for indirect emissions from leaching of N from soils. Equation 3.9 provides a simplified equation for calculating indirect N_2O emissions via atmospheric deposition from land application of wastewater treatment plant sludge, while Eq. 3.10 is a simplified equation for calculating indirect N_2O emissions via leaching.

$$N_2O_{\text{Indirect atm}} = F_{\text{SEW}} \times F_{\text{GASM}} \times EF \times \frac{44}{28} \tag{3.9}$$

where

$N_2O_{\text{Indirect atm}}$ = annual indirect N_2O emissions produced via atmospheric deposition of N volatilized from managed soils in inventory year (kg N_2O/yr);

F_{SEW} = annual amount of nitrogen in sewage sludge applied to land (kg N/yr);

F_{GASM} = fraction of applied N in sewage sludge that volatilizes as NH_3 and NO_x (kg N volatilized/kg N applied); and

EF = emission factor for N_2O emissions from atmospheric deposition of N on soils and water surfaces (kg N_2O–N/kg NH_3–N and NO_x volatilized).

$$N_2O_{\text{L}} = F_{\text{SEW}} \times F_{\text{Leach}} \times EF \times \frac{44}{28} \tag{3.10}$$

where

N_2O_L = annual indirect N_2O emissions due to leaching of N from managed soils in inventory year (kg N_2O/yr);

F_{Leach} = fraction of applied N in sewage sludge that leaches as NH_3 and NO_x (kg N leached or lost via runoff/kg N applied); and

EF = emission factor for N_2O emissions from N leaching and runoff (kg N_2O–N/kg N leached or lost via runoff).

The IPCC provides default factors of 0.2 and 0.3 for the amount of N applied that is volatilized and leached (i.e., F_{GASM} and F_{Leach} respectively), respectively. The IPCC also provides emission factors of 0.01 kg N_2O–N/kg N and 0.0075 kg N_2O–N/kg N for the calculation of N_2O emissions from volatilized and leached N, respectively. Countries are encouraged to determine country-specific leaching, volatilizing and emission factors.

3.5.2 Incineration

Incineration remains a common form of sludge disposal in many developed countries. Modern incineration facilities may operate with a net positive energy footprint. This affects national GHG emissions from energy generation, therefore the IPCC accounts for this in the guidelines for GHG emissions from energy generation. The direct GHG emission from incineration of sludge is estimated based on the organic content of the sludge, and the fraction of organic carbon which may be oxidized to and converted to GHGs. CO_2, CH_4 and N_2O are generated during incineration. For CO_2 emission, Tier 1 and 2 methodologies are based on Eq. 3.11, with Tier 1 using default emission factors and Tier 2 applying country-specific emission factors.

$$CO_{2,Emission} = SW \times dm \times CF \times FCF \times OF \times \frac{44}{12} \tag{3.11}$$

where

$CO_{2,Emission}$ = CO_2 emissions in inventory year (Gg/yr);

SW = total amount of sludge (wet weight) incinerated or open-burned (Gg/yr);

dm = dry matter content in the sludge (wet weight) incinerated or open-burned;

CF = fraction of carbon in the dry matter (total carbon content);

FCF = fraction of fossil carbon in the total carbon; and

OF = oxidation factor.

44/12 is used to convert C to CO_2 since the atomic mass and the relative molecular mass of C and CO_2 are 12 and 44, respectively.

Since carbon in the wastewater, and consequently in the sludge, is biogenic, so the default emission factors provided by the IPCC for sewage sludge are *CF* of 40–50, *FCF* of 0 and *OF* of 100. For Tier 3 estimation, incinerator plant-specific data are required to compile a bottom-up estimation of GHG emissions. While the GHG generated from incineration will be primarily comprised of CO_2 emissions, CH_4 emissions may occur from incomplete combustion, and N_2O emissions may result from combustion of nitrogenous waste. Tier 1 and 2 CH_4 emissions methodologies are based on Eq. 3.12, with Tier 1 using default emission factors and Tier 2 applying country-specific emission factors.

$$CH_{4,\text{Emission}} = SW \times EF \times 10^{-6} \tag{3.12}$$

where

EF = aggregate CH_4 emission factor (wet weight) (kg CH_4/Gg of waste); and
10^{-6} = conversion factor from kilograms to gigagrams.

The CH_4 emission factor is dependent on the incineration technology used. The IPCC does not provide specific CH_4 emission factors for sludge incineration. Again, incinerator plants' specific data are required to compile a Tier 3 bottom-up estimation of GHG emissions. N_2O emissions are determined using Eq. 3.13 for Tier 1 and Tier 2. N_2O emissions from incineration are dependent on the type of incineration systems.

$$N_2O_{\text{Emission}} = SW \times EF \times 10^{-6} \tag{3.13}$$

where

EF = N_2O emission factor for sludge (wet weight) (kg N_2O/Gg of waste).

Tier 1 and 2 N_2O emission methodologies are based on Eq. 3.13, with Tier 1 using default emission factors and Tier 2 applying country-specific emission factors to the variables outlined. The default factor for sewage sludge provided by the IPCC for N_2O emissions from incineration is 0.9 kg N_2O/Gg.

3.5.3 Landfill

Despite regulatory pressure to reduce the amount of biodegradable waste entering landfill, landfilling of sewage sludge remains a popular low-cost sludge disposal option. It should be noted that even though CO_2 is the GHG emitted in the greatest quantities in landfills, as this CO_2 is of biogenic origin, it is not accounted for in national GHG emission estimates. As with other IPCC guidelines, Tier 1 guidelines utilize default emission factors while Tiers 2 and 3 require increasingly detailed country-specific data to determine total GHG emissions from landfilling.

The IPCC guidelines for determining the GHG emissions from landfilled sludge are based upon assuming a first order decay relationship of carbon stored in landfilled sludge converting to GHGs (specifically CH_4). Such a relationship means that when a waste is landfilled the rate of CH_4 emission is directly related to the amount of degradable organic carbon (DOC) in the waste.

As CH_4 emissions are dependent on the total amount of biodegradable carbon in the landfill (rather than annual deposits of waste), determining GHG emissions from national landfills attributable specifically to sewage sludge requires data on the amounts of sewage sludge disposed to landfill for an extended period of time (the IPCC suggests a minimum of 50 years). Equation 3.14 is used to determine annual CH_4 emissions from landfills.

$$\mathrm{CH}_{4,\mathrm{Emission}} = \left(\sum_i \mathrm{CH}_{4,\mathrm{generated},i,T} - R_T\right) \times (1 - OX_T) \tag{3.14}$$

where

T = inventory year;

i = type/category of waste, among which sludge is one since sludge is landfilled with other wastes;

R_T = recovered CH_4 in year T (Gg); and

OX_T = oxidation factor in year T.

The oxidation factor in the above equation is the fraction of CH_4 generated in landfills which is subsequently oxidized to CO_2 and water in the landfill prior to emission or flaring. The capturing and utilization of landfill gas for energy generation is common in many countries. This affects national GHG emissions from energy generation, therefore the IPCC accounts for this in the guidelines for GHG emissions from energy generation. The R_T value accounts for the avoidance of direct emissions due to CH_4 utilization.

It is clear that Eq. 3.14 requires an estimation of the gross CH_4 generation from landfills. This is highly dependent on the amount of decomposable degradable organic carbon (DDOC) in the waste. Degradable organic matter may not undergo decomposition due to poor accessibility of waste by microorganism, inhibition of bacteria due to toxicity of wastes and intermediates etc. Figure 3.4 shows the mass balance of degradable organic carbon in the landfill site.

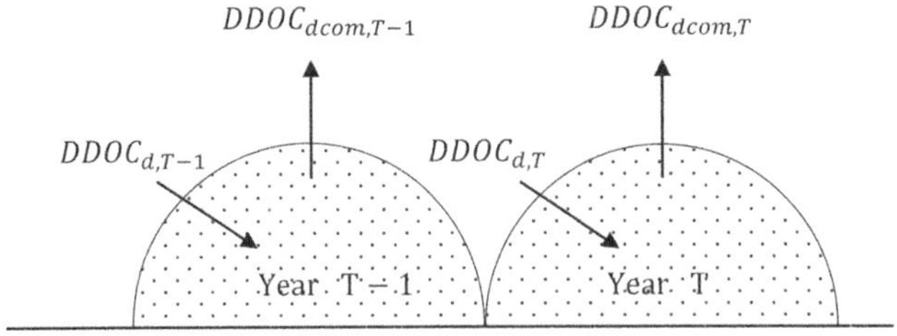

Figure 3.4 Mass balance of degradable organic carbon in the landfill site.

Decomposable DDOC at end of the inventory year T is estimated using Eq. 3.15. It is determined as the amount of DDOC accumulated in the landfill from the waste in question at the end of the target year, and the rate at which CH_4 is emitted for the waste in question (in terms of first order reaction constant).

$$DDOC_{dcom,T} = DDOC_{a,T-1} \times (1 - e^{-k}) \tag{3.15}$$

where

$DDOC_{a,T-1}$ = DDOC accumulated in the landfill sites at the end of year $T - 1$ (Gg);
$DDOC_{dcom,T}$ = DDOC decomposition in the landfill sites in year T (Gg);
k = reaction constant, $k = \ln 2/t_{1/2}$ (yr^{-1}); and
$t_{1/2}$ = half-life time (yr).

$DDOC_{a,T-1}$ is DDOC accumulated in the landfill sites at the end of year $T - 1$ (Gg). It is calculated with Eq. 3.16:

$$DDOC_{a,T} = DDOC_{d,T} + DDOC_{a,T-1} \times e^{-k} \tag{3.16}$$

where

$DDOC_{d,T}$ = DDOC deposited into the landfill sites in year T (Gg).

The IPCC provides default factors for sewage sludge for the reaction constants for a range of different climates (which have a major effect on the rate of methanogenic activity) in the range of 0.06 to 0.7 yr^{-1}. For Tier 2 or Tier 3 calculations, these values should be derived experimentally. Equation 3.17 is the IPCC guideline for determining gross CH_4 generations from landfills per year.

$$CH_{4,Generation,T} = DDOC_{dcom,T} \times F \times \frac{16}{12} \tag{3.17}$$

where

$CH_{4,Generation,T}$ = amount of CH_4 generated from decomposable material in inventory year T;
F = fraction of CH_4 by volume, in generated landfill gas (%);
16/12 is used to convert C to CH_4.

3.5.4 Composting

The composting of wastewater treatment sludge may be undertaken on dewatered sludge and on dewatered digestate generated from anaerobic digestion of sludge. While most jurisdictions allow for the land application of anaerobically digested sludge directly to land, many countries require sewage sludge to meet hygiene standards prior to land application. Composting is a common means of providing further sanitation of wastewater treatment sludge prior to land application.

Composting generates CO_2 and N_2O emissions and to a lesser extent, CH_4. As CO_2 is of biogenic origin, it is not included in the estimations made using the IPCC guidelines. Rather than provide specific methodology for the estimation of GHG emission from composting specifically, the IPCC provides generalized guidelines for the estimation of GHG emission from biological waste treatment methods.

In a similar manner to other section of the guidelines, Tier 1 methodology provides default factors for the variables in the following equations (Eqs. 3.18 and 3.9), while Tier 2 methodology requires country-specific factors to be used. Tier 3 recommends a bottom-up approach to determining GHG emission. The two equations provided for the determination of annual GHG emissions from composting are similar, with the mass of waste composted and an emission factor determining the amount of GHG released.

$$CH_{4,\mathrm{Emission}} = M \times EF \times 10^{-3} - R \tag{3.18}$$

where

M = amount of sludge composted in inventory year (Gg);
EF = emission factor (kg CH_4/kg sludge composted); and
R = total amount of CH_4 recovered in inventory year (Gg CH_4).

$$\mathrm{N_2O}_{\mathrm{Emission}} = M \times EF \times 10^{-3} \tag{3.19}$$

where

EF = emission factor (kg N_2O/kg sludge composted).

The default emission factors provided by the IPCC for CH_4 and N_2O emissions are 4 g CH_4 and 0.3 g N_2O/kg sludge (wet weight), illustrating that CH_4 emission from composting is typically of a greater magnitude than N_2O emission. However, after the global warming potential (GWP) of each of these gases (21 for CH_4 and 265 for N_2O) is considered it appears that N_2O emissions are the greatest contributing factor to GHG emissions from composting.

3.6 UNCERTAINTY

IPCC inventory calculations, which rely on the estimations of population, types of wastewater treatment systems and the associated population, wastewater characteristics (estimation of influent BOD_5, COD and N concentrations), emission factors, etc., are deemed to have a high degree of uncertainty. Every country needs to decide the appropriate methods and parameters to estimate their emissions inventory, depending on the availability of data. Herein, the methodologies used by two countries, the USA and Ireland, to calculate their national GHG emissions inventory for wastewater treatment systems are compared. The comparison shows uncertainties associated with IPCC methodology, and the importance of conducting intensive field research to acquire data for estimating the emissions of these gases.

3.6.1 CH_4 emissions from domestic wastewater

The Irish Environmental Protection Agency (EPA) splits the domestic wastewater treatment systems into three types: septic tanks; centralized wastewater treatment with biogas recovery; and centralized wastewater treatment without biogas recovery (so $j = 3$ in Eq. 3.1) (Duffy *et al.* 2015).

TOW is calculated for the whole country, described in Eq. 3.20, based on the population equivalent data and allocating among the three wastewater treatment systems.

$$TOW = PE \times \text{BOD} \times 0.001 \times I \times 365 \tag{3.20}$$

where

BOD = per capita BOD_5 in inventory year (g/person/day), 60 g/person/day is used;

0.001 = conversion from grams to kilograms; and

I = correction factor for additional industrial and commercial BOD discharged to sewers, I is equal to 1.25.

For *EF*, the Irish EPA uses country-specific data. Because the temperature suitable for methanogenesis occurring in septic tanks is found in only two summer months, *MCF* is assumed to be lower than the IPCC default value, 0.083. So, *EF* for a septic tank is 0.05 kg CH_4/kg BOD_5. For centralized wastewater treatment without biogas recovery, it is assumed that 3% wastewater is subjected to anaerobic conditions ($MCF = 0.03$), so $EF = 0.018$ kg CH_4/kg BOD_5. A total of 45% of BOD loading in Ireland is treated in centralized WWTPs with biogas recovery, so the Irish EPA thinks that CH_4 emission from this category is zero (which is actually not the case, as discussed in Chapter 6).

The United States Environmental Protection Agency (USEPA) splits the wastewater treatment systems into seven categories: (1) septic tank systems; (2) well managed aerobic wastewater treatment systems; (3) not well managed wastewater treatment systems with primary treatment; (4) not well managed wastewater treatment systems without primary treatment; (5) centrally treated anaerobic systems with primary treatment; (6) centrally treated anaerobic systems without primary treatment; and (7) centrally treated anaerobic systems with biogas recovery (USEPA, 2015).

When estimating septic tank system emissions, USEPA uses Eq. 3.21, rather than Eq. 3.1:

$$\text{CH}_{4,\text{Emission}} = PE \times (\%\text{onsite}) \times EF_{\text{septic}} \times 10^{-9} \times 365.25 \tag{3.21}$$

where

%onsite is the flow to septic tank systems divided by the total flow, and

EF_{septic}, EF of septic systems, is 10.7 g CH_4/capita/yr.

USEPA uses Eq. 3.1 to estimate CH_4 emission from aerobic and anaerobic wastewater treatment systems. For well managed aerobic systems, $EF = 0$, but for not well managed aerobic systems, default $MCF = 0.3$ is used (which is much higher than 0.03 used by the Irish EPA), so $EF = 0.18$ kg CH_4/kg BOD. For centralized anaerobic digestion system, the default MCF is used, so $EF = 0.48$ kg CH_4/kg BOD. When estimating CH_4 emission from anaerobic systems with biogas recovery, USEPA has a different attitude to the Irish EPA. It assumes that 1% of biogas produced is emitted to the atmosphere. Therefore, the emissions from anaerobic digesters are calculated as:

$$CH_{4,\text{Emission}} = Q_{AD} \times Y_{\text{biogas},PE} / Q_{PE} \times F \times \rho \times (1 - R) \times 10^{-9} \times 365.25 \quad (3.22)$$

where

Q_{AD} = wastewater influent flow to WWTPs with biogas recovery (m^3/day);

$Y_{\text{biogas},PE}$ = biogas produced from organic matter in wastewater per capita per day (m^3/capita/day);

Q_{PE} = per capita wastewater flow per day (m^3/capita/day);

F = fraction of CH_4 in biogas; and

ρ = methane density.

USEPA uses Q_{PE} of 378 litre/person/day (100 gal/person/day) and F of 65% in the estimation.

From the calculations, it is found that septic tanks contributed to 65.5% of CH_4 emissions from domestic wastewater treatment systems in the USA in 2014 and centralized wastewater treatment systems contributed to 34.5%.

3.6.2 CH_4 emissions from industrial wastewater

The Irish EPA did not report CH_4 emissions from industrial wastewater separately; CH_4 emissions from industrial wastewater were considered when calculating TOW with an industrial factor, I (Eq. 3.19). However, USEPA took into account CH_4 emissions from five major industrial wastewater categories: meat and poultry; pulp and paper; fruit and vegetables; petroleum refineries; and ethanol production, because of their high organic loadings. The total CH_4 emissions from industrial wastewater are the sum of the emissions from these five industries.

3.6.3 N_2O emissions from wastewater treatment facilities

The Irish EPA ignores direct N_2O emission from wastewater treatment facilities, during the estimation of national GHG emission inventory. USEPA estimates it using Eq. 3.6 by breaking down the wastewater treatment systems into (1) septic systems where N_2O emission is zero because of no nitrification/denitrification occurring; (2) centralized WWTPs with nitrification/denitrification occurring; and (3) centralized WWTPs without nitrification/denitrification occurring. The *EF* used are USA specific data obtained through local research: (1) *EF* for plant with intentional denitrification: 7 g N_2O/capita/yr; and (2) *EF* for plant without intentional denitrification: 3.2 g N_2O/capita/yr (Scheehle & Doorn, 2001).

USEPA and the Irish EPA both use Eq. 3.4 to estimate N_2O emission from wastewater effluent. The parameters F_{NPR}, $F_{Non\text{-}Con}$, $F_{Ind\text{-}Com}$, and $EF_{Effluent}$ are all the IPCC default values, while the per capita protein consumptions are country-specific data: 29.5–31.9 kg/capita/year in the USA and 39.4–43.1 kg/capita/year in Ireland.

The above comparison clearly shows that in order to acquire accurate estimates, intensive field research needs to be carried out to identify country-specific data. These CH_4 and N_2O emission factors used by both countries and the default values provided by IPCC guidelines are based on a limited number of research results, which is not sufficient due to the complexity of wastewater treatment systems. This will be discussed in Chapters 5 and 6.

3.6.4 Source of uncertainties

The different approaches applied in the USA and Ireland reveals that there are great uncertainties in estimation of GHG emissions. These uncertainties can be related to, if we ignore the uncertainties associated with the errors of measurement of direct N_2O and CH_4 emissions in field studies due to instruments, people and testing methodologies: (1) uncertainties associated with method selection, i.e., selection of mathematical equations or suitable Tiers (Tiers 1, 2 or 3); (2) uncertainties associated with parameters, i.e., selection of emission factors and values of other parameters in equations; and (3) uncertainties associated with activity data, i.e., data of key categories. These uncertainties are due to our limited knowledge of GHG emissions from wastewater treatment facilities and limited data available. Activity data are closely linked to economic activity, which are collected by the national statistical offices regularly. So, the uncertainties associated with the activity data are much less than the uncertainties associated with the selection of emission factors. After the methods are decided, determination of emission factors will be very important since the emission factors measured in field studies vary a lot. Table 3.3 lists N_2O emission factors for USA WWTPs.

Table 3.3 shows that the emission factors vary plant by plant (for the same country). Even for the same plant, the emission factors measured in different sample rounds vary significantly. Hence, to use one emission factor value acquired in a single plant and in one single testing campaign as the emission factor for WWTPs in the whole country brings about significant uncertainties. Therefore, the IPCC has prepared Good Practice Guidance and Uncertainty Management in

National Greenhouse Gas Inventories for dealing with the uncertainties (Penman *et al.* 2000) and Chapter 6, Volume 1 of its 2006 Guidelines for National GHG Inventories details how to carry out quality assurance and quality control and verification (IPCC, 2006). To reduce the uncertainties, fundamentally, we need to carry out intensive field testing so as to improve our understanding of GHG emissions and mitigation in WWTPs and to increase the data availability.

Table 3.3 N_2O emission factors surveyed.

WWTP	Emission Factor (kg N_2O-N/kg TKN[a] Influent)
Separate-stage BNR[b] process	0.0001–0.0003[c]
Four-stage Bardenpho process	0.0016–0.006
Step-feed BNR process – Plant 1	0.0062–0.016
Step-feed BNR process – Plant 2	0.015
Step-feed BNR process – Plant 3	0.0005
Step-feed non-BNR process	0.0018–0.018
Separate centrate	0.0024–0.0054
Plug-flow process – Plant 1	0.004
Plug-flow process – Plant 2	0.0009–0.0062
MLE[d] process – Plant 1	0.0007
MLE process – Plant 2	0.0006
Oxidation ditch process	0.0003

[a]TKN = Total Kjeldahl Nitrogen; [b]BNR = Biological Nitrogen Removal; [c]indicates emission factors measured in different sampling campaigns in the same WWTP; [d]MLE = Modified Ludzack-Ettinger.
Source: Foley *et al.* (2011).

3.7 SUMMARY

GHG emissions from wastewater treatment systems are a small proportion of the national GHG emission inventories, for instance, only 0.3% in American and Irish national GHG emissions in 2013 (Duffy *et al.* 2015; USEPA *et al.* 2015). However, with the increase in the population, the increase in the protein content in the human diet and the increased use of advanced wastewater treatment facilities for nitrogen removal, there is a trend that the GHG emissions from wastewater treatment sector are increasing. Due to the limited available data for CH_4 and N_2O emissions from wastewater treatment facilities, the default values applied by the IPCC may underestimate CH_4 and N_2O emissions by one order of magnitude due to their lack of scientific basis or due to the less systematic data used (López *et al.* 2013). Therefore, it is necessary to further examine CH_4 and N_2O emissions from various sources during wastewater treatment.

3.8 REFERENCES

Czepiel P., Crill P. and Harriss R. (1995). Nitrous oxide emissions from municipal wastewater treatment. *Advances in Environmental Science and Technology*, **29**(9), 2352–2356.

Duffy P., Hanley E., Black K., O'Brien P., Hyde B., Ponzi J. and Alam S. (2015). Ireland National Inventory Report 2015: Greenhouse Gas Emissions 1990–2013 Reported to the United Nations Framework Convention on Climate Change Environmental Protection Agency. Irish EPA, 2015.

Foley J., Yuan Z., Keller J., Senante E., Chandran K., Willis J., Shah A., van Loosdrecht M. and van Voorthuizen E. (2011). N_2O and CH_4 Emission from Wastewater Collection and Treatment Systems: Technical Report. Global Water Research Coalition, London, United Kingdom.

Hobson J. (2000). CH_4 and N_2O emissions from waste water handling. Good practice guidance and uncertainty management in National Greenhouse Gas Inventories. Intergovernmental Panel on Climate Change (IPCC) Publications, Geneva, Switzerland.

IPCC (2006). 2006 IPCC guidelines for national greenhouse gas inventories. Figure 6.1, Figure 6.2 and Table 6.3 in Chapter 6, Volume 5. Prepared by the National Greenhouse Gas Inventories Programme, H. S. Eggleston, L. Buendia, K. Miwa, T. Ngara and K. Tanabe (eds), Institute for Global Environmental Strategies, Kanagawa, Japan.

López J. C., Quijano G., Souza T. S., Estrada J. M., Lebrero R. and Muñoz R. (2013). Biotechnologies for greenhouse gases (CH_4, N_2O, and CO_2) abatement: state of the art and challenges. *Applied Microbiology and Biotechnology*, **97**(6), 2277–2303.

Penman J., Kruger D., Galbally I. E., Hiraishi T., Nyenzi B., Emmanuel S., Buendia L., Hoppaus R., Martinsen T., Meijer J., Miwa K. and Tanabe K. (2000). Good Practice Guidance and Uncertainty Management in National Greenhouse Gas Inventories. Intergovernmental Panel on Climate Change, Kanagawa, Japan.

Scheehle E. A. and Doorn M. R. (2001). Improvements to the US Wastewater Methane and Nitrous Oxide Emissions Estimates. USEPA, Washington, DC.

United States Environmental Protection Agency (2015). Inventory of U.S. Greenhouse Gas Emissions and Sinks: 1990–2013.

Chapter 4

Measurement of direct greenhouse gas emissions in WWTPs

Xinmin Zhan

4.1 INTRODUCTION

If we want to build future zero-carbon footprint wastewater treatment plants (WWTPs), it is important to know the quantity of greenhouse gases (GHGs) emitted directly from wastewater treatment facilities. Measurement of direct GHGs emissions is not easy due to the changing features of influent wastewater characteristics (seasonal and diurnal variations), climate, operation strategies and system performance, and the complexity in setting up testing protocols. In this chapter, the techniques that have been used in a limited number of studies are briefly introduced. These techniques are applied to measure the off-gas from WWTPs or measure dissolved gas concentrations in the mixed liquor.

4.2 OFF-GAS MEASUREMENT

When measuring the off-gas, there are no common standards available, because the wastewater treatment units are so different. Some units are aerated, such as aerated activated sludge tanks, and some are not aerated like clarifiers, anoxic tanks and anaerobic tanks. Some wastewater tanks are covered but a large amount of tanks are not. The testing techniques must be developed to accommodate individual situations.

The principle of quantification of gas emissions by carrying out off-gas measurement is to measure the gas flow rate and the concentrations of targeted gases (CO_2, CH_4, N_2O) in the off-gas:

$$Q_{E,i} = \frac{Q_{oG} C_i M_i}{22.4 \times 10^6} \tag{4.1}$$

Xinmin Zhan, Zhenhu Hu, Guangxue Wu
doi: 10.2166/9781780406312_053

where

$Q_{E,I}$ = amount of gas i (CO_2, CH_4, or N_2O) emission (kg/day);
C_i = concentration of the gas i in the off-gas measured (ppmv);
Q_{oG} = flow rate of the off-gas at the standard temperature and pressure (STP) (m^3/day);
M_i = the molecular mass of gas i (g/mol); and
i = CO_2, CH_4 or N_2O.

The emission factors of CH_4 and N_2O for individual wastewater treatment facilities are calculated with Eq. 4.2:

$$EF = \frac{Q_{E,i}}{Q_I C_{I,i}} \tag{4.2}$$

where

Q_I = influent wastewater flow rate (m^3/day); and
$C_{I,i}$ = concentration of contaminants in the influent corresponding to gas i (kg/m^3). For the emission factor of CH_4, $C_{I,i}$ is the concentration of chemical oxygen demand (COD) in the influent; for the emission factor of N_2O, $C_{I,i}$ is the concentration of influent N (in most cases, this is NH_4^+-N).

The concentrations of gases in the off-gas are measured using gas chromatograph or infrared methods.

4.2.1 Enclosed wastewater treatment facilities

Some local authorities require WWTPs to be enclosed or certain wastewater treatment tanks to be covered so as to prevent the release of malodors and noise to the surrounding environment, and reduce local resident complaints. Covering the plants wholly or partly may also reduce adverse impacts on the landscape. The off-gas from the covered WWTPs and tanks will be collected and then discharged to the air via ventilation pipes after the off-gas is treated. Figure 4.1 shows the covered screening and primary clarifiers in the Reading Wastewater Treatment Plant, United Kingdom, as well the ventilation pipe which discharges the off-gas to the air after the gas is treated for removing H_2S.

If the tanks or the whole plants are covered, the off-gas flow rate can be measured by testing the gas flow velocity in the ventilation pipes, and the concentrations of GHGs in the off-gas can be measured with infrared techniques. One significant advantage of measurement of off-gas in covered tanks is that the total gas emissions can be monitored continuously, in addition to the temporal variations. When the whole WWTP is covered, the gas emission measured is for the whole plant and there is no distinction of gas emissions from individual process units. This means that the spatial variations of GHG emissions in the plant cannot be identified.

Figure 4.1 Covered screening and primary clarifier in Reading WWTP (a) inside the building where vents are used to collect off-gas; (b) the ventilation pipe beside the building.

In order to measure the flow rate of the off-gas, Q_{oG}, pitot tubes are installed on the ventilation pipe. They measure the gas velocity based on the pressure difference between the total pressure (static pressure plus the dynamic pressure) and the static pressure (Figure 4.2). After knowing the pressure difference, the gas velocity is calculated as:

$$u = \sqrt{\frac{2\Delta P}{\rho}} \tag{4.3}$$

where

u = gas flow velocity (m/s);
ΔP = difference between the total pressure and the static pressure (Pa); and
ρ = density of the gas at the local conditions (kg/m^3).

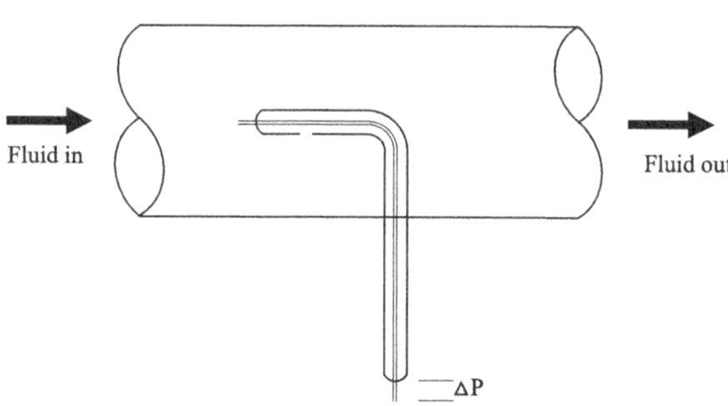

Figure 4.2 Working mechanism of pitot tubes.

For the accuracy of measurement, the pitot tubes should measure the gas flow velocities at 20–50 locations along the ventilation pipe (Foley *et al.* 2011). The pitot tubes are also able to measure temperature, which will help the conversion of gas flow velocity at certain temperatures to STP conditions.

Q_{OG} is calculated with Eq. 4.4 based on the internal diameter of the ventilation pipe and the gas flow velocity measured:

$$Q_{OG} = \frac{1}{4}\pi D^2 u \times 3600 \times 24 \tag{4.4}$$

where

D = internal diameter of the ventilation pipes (m).

The value 3600 is used to convert the units from m^3/s to m^3/h and 24 converts m^3/h to m^3/day.

Alternatively, the gas flow velocity can also be measured using anemometers at pipe outlets, which are widely used in weather stations to measure wind velocity. Researchers in the Netherlands have tested GHG emission factors of three enclosed wastewater treatment plants: Papendrecht, Kralingseveer and Kortenoord (Foley *et al.* 2011). Pitot tubes were used on Papendrecht and Kralingseveer WWTPs to measure the off-gas flow velocity and a hot wire anemometer was applied in Kortenoord WWTP.

4.2.2 Off-gas measurement techniques for open aerated tanks

The simplest method for monitoring off-gas from open aerated tanks is the bag technique, which was used by Czepile *et al.* (1993) to measure CH_4 and CO_2 fluxes. A polyethylene sample bag is fastened to the inside of a wood support frame (Figure 4.3) and a thermocouple thermometer is mounted inside the bag to record the temperature. When the testing commences, the ambient air in the polyethylene bag is emptied, and then the wood support frame (in addition with the bag held) is immersed several centimetres deep into the wastewater of the tested surface of the aerated tank. After the bag is filled with gas, gas samples are taken from the bag with syringes and gas composition is determined using gas chromatography (GC). The gas emission flux per area of the water surface for gas i, $F_{E,i}$ ($kg/m^2 \cdot day$), is calculated using Eq. 4.5 (Czepiel *et al.* 1993).

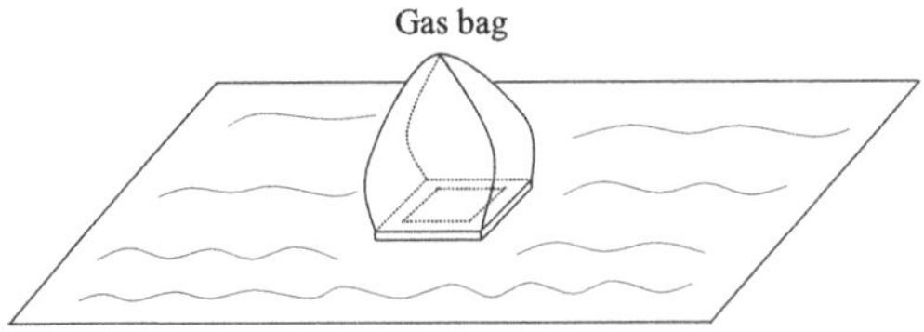

Figure 4.3 Bag sampling technique for GHG emission.

$$F_{E,i} = \rho_i \times c_i \times Q/A \tag{4.5}$$

where

ρ_i = density of the gas *i* at the temperature recorded in the sample bag (kg/m^3);
c = gas *i* concentration in the bag (m^3/m^3);
Q = diffused air flow (m^3/day); and
A = the total surface area of the aerated tank.

$$Q_{E,i} = F_{E,i} \times A \tag{4.6}$$

In this method, the air blowers' air flow rate for the aerated tank is estimated as the off-gas flow rate. This is also used by other researchers (Aboobakar *et al.* 2014). However, it is not accurate to use the blowers' air flow rate to represent the off-gas flow rate because of (1) the difference between the pressure and temperature of compressed air and the off-gas; (2) the generation of gases in the aeration tanks; and (3) the uneven distribution of the air flow along the tank surface. However, this technique is simple, cheap, and easy to execute in a field study. Another advantage of this technique is that it combines the gas concentration data with the tank's operating conditions (air flow rate).

Other practical testing protocols are to have floating hoods or flux-chambers suspended on the aerated tank surface. After measurement of the off-gas flow rate and the concentration of gases in the off-gas, the GHG emission from the surface covered by the floating hoods or flux-chambers is calculated using Eq. 4.1, and then the total GHG emission from the whole aeration tank is calculated by assuming that the gas emission flux is uniform over the whole surface area of the aeration tank:

$$Q_{E,i} = \frac{Q_{E,i,\text{flux chamber}}}{A_{\text{flux-chamber}}} A_S \tag{4.7}$$

where

$A_{\text{flux-chamber}}$ = surface area of the floating hoods or flux-chambers (m^2);
A_s = surface area of the aeration tank (m^2); and
$Q_{E,i,\text{ flux chamber}}$ = emission flux of gas *i* measured with the flux chamber (kg/day).

Figures 4.4 and 4.5 show the floating hood system used by Pan *et al.* (2016). The floating hoods, which are made of plastic materials, are suspended on the water surface and attached to tank skirts for position. They are submerged below the water surface to 100–150 mm. The off-gas flow rate is continuously measured with gas flow meters which are set up in the exhaust gas pipe. The concentrations of gases are measured using the infrared method.

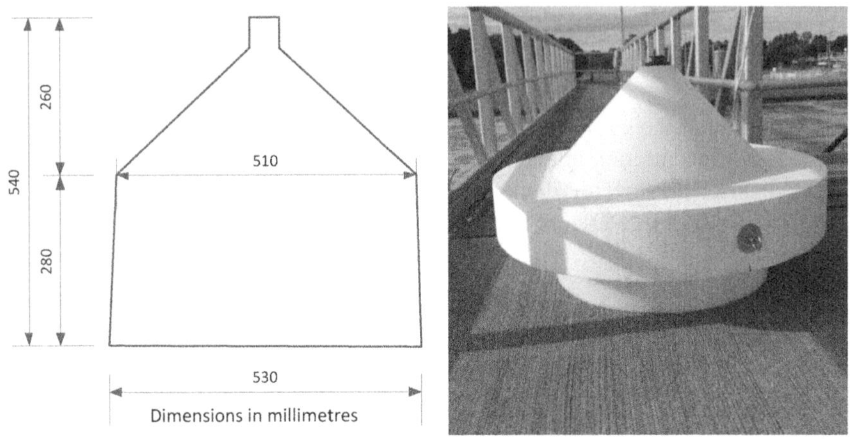

Figure 4.4 Floating hood used in the aeration tanks (Pan *et al.* 2016).

One problem with the floating hood method is that it is difficult to obtain steady advective gas flow rate measurement, which can be solved using the flux-chamber method. A widely used flux-chamber method is the surface emission isolation flux chamber (SEIFC) method recommended by the United States Environmental Protection Agency (USEPA) which is based on USEPA/600/8-86/008 and the South Coast Air Quality Management District (SCAQMD) tracer method. A schematic diagram of a SEIFC flux chamber is shown in Figure 4.6 (Chandran *et al.* 2016). To obtain a uniform gas composition in the chamber when measuring the off-gas flow rate, mixing through a mechanical mixer or through sweep gas circulation is provided. When the SEIFC method is applied to the aerated tanks, sweep gas circulation is not needed. When carrying out the testing, the flux chamber is lowered in the area of the aeration tank and its bottom is below the water surface by 250–500 mm. When measuring the off-gas flow rate, a tracer gas, commonly using helium gas (10% helium and 90% zero air), is introduced into the chamber at a known flow rate (1 L/min). The concentration of the tracer gas in the exhaust gas from the chamber is measured. Then, the off-gas flow rate is calculated using Eq. 4.9 based on the mass balance of helium gas in the SEIFC chamber (shown in Figure 4.7):

$$Q_{\text{tracer}} C_{\text{He-tracer}} = (Q_{\text{tracer}} + Q_{OG}) \text{C}_{\text{He-OG}} \tag{4.8}$$

$$Q_{OG} = \frac{Q_{\text{tracer}} (C_{\text{He-tracer}} - C_{\text{He-OG}})}{C_{\text{He-OG}}} \tag{4.9}$$

in which, $C_{\text{He-tracer}}$ and $C_{\text{He-OG}}$ are concentrations of the tracer gas in the input tracer gas and the exhaust gas, respectively. Q_{tracer} is the tracer gas' input flow rate.

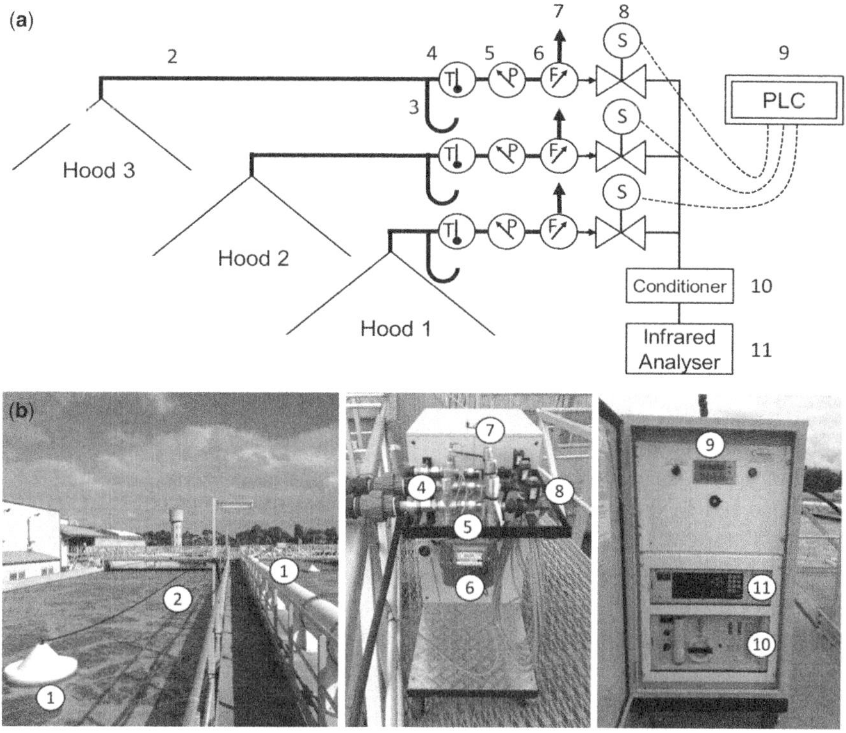

Figure 4.5 Floating hood system in the field study: (a) line diagram; (b) photograph of the set-up. (1) gas hoods positioned along the tank surface; (2) polyethylene gas tubing; (3) condensation moisture trap; (4) temperature sensor; (5) pressure sensor; (6) flow meter; (7) excess gas exhaust; (8) solenoid gas valves for multiple hood gas sampling; (9) programmable logic controller; (10) gas sample conditioning system; (11) infrared N_2O, CH_4, and O_2 gas analyzer (Pan *et al.* 2016).

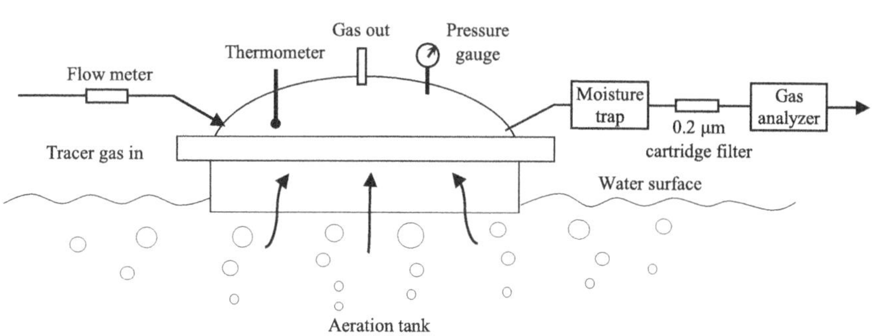

Figure 4.6 USEPA's surface emission isolation flux chamber.

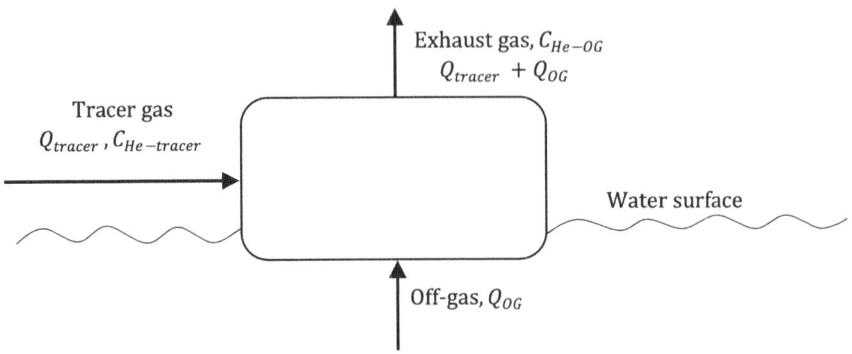

Figure 4.7 Tracer gas mass balance for SEIFC flux chamber.

When measuring the off-gas composition, the on-line N_2O and CH_4 analyzers are connected to the flux chamber's gas testing vents. The off-gas flow rate is measured discretely and the gas composition is able to be continuously measured. Hence, it is important to set up the flux chamber location and testing time properly so as to obtain the representative off-gas flow rates. Measurements should be taken in at least two locations in the targeted aeration tank.

4.2.3 Measurement of off-gas in non-aerated tanks

4.2.3.1 Static chamber technique

In the unit processes where there is no aeration provided (anoxic tank, anaerobic tank, sludge conditioning tank, clarifiers, etc.), GHG emission is mainly due to diffusion. A static chamber flux measurement methodology can be used (Mosier *et al.* 1996). The chambers can be constructed using plastic (like polyethylene) pipes (Figure 4.8). A 10-mm hole is drilled in the chamber lid and a rubber septum stopper is placed in the hole. A thermometer is set up in the chamber to measure the temperature in the headspace. A sampling port is created in the lid of the chamber using a connector, semi rigid nylon tube and a syringe. A 3-way valve is fitted to allow for accurate sampling. The static chamber is held to a float (for example a wood frame) and then anchored in the unit process tanks by attaching it to tank skirts for position with the bottom of the rim of the chamber submerged in the water at approximately 100 mm depth. When in place, the depth from the water surface of the tank to the chamber lid is measured, H (m).

When the testing commences, the chamber is anchored in position. Ten-millilitre gas samples are taken at time 0, and then at 2-min intervals for 10 minutes after placing the chamber. The corresponding temperatures are also recorded. The composition of gas samples is measured using GC.

Figure 4.8 Static flux chamber used for measurement of gas flux from pilot scale woodchip biofilters (Carney, 2012).

The fluxes are calculated from the rate of concentration change of the GHGs (N_2O, CH_4 or CO_2) in the chamber headspace according to Eq. 4.10 (Hutchinson & Mosier, 1981; Carney, 2012).

$$F_{E,i} = H \times \frac{dC_i}{dt} \times \frac{P}{RT} \times M_i \times 60 \times 24 \tag{4.10}$$

where

E = emission flux (kg/m^2/day);
C_i = concentration of gas i ($N_2O/CO_2/CH_4$) in gas in the headspace (ppmv);
t = time for sampling, min;
P = atmospheric pressure measured in the field (atm);
R = gas constant (0.082 atm L K^{-1} mol^{-1});
T = gas temperature measured in the chamber (K);
M_i = molecular mass of gas i, e.g., N_2O 44 g/mole, CH_4 16 g/mole and CO_2 44 g/mole;
dc/dt = linear increase in gas concentrations in the chamber during the sampling period; and 60 and 24 are used to convert minutes to days.

A battery-operated blower can be put in the chamber to enable a good mixing of gas inside the chamber. The static chamber technique is the simplest design and is very easy to set up on site.

Using this approach, one problem is that there might be insufficient mixing in the chamber, so the samples may not be representative. Therefore, an improved design, as shown in Figure 4.9, is to provide a sweep gas to the system so as to create sufficient mixing.

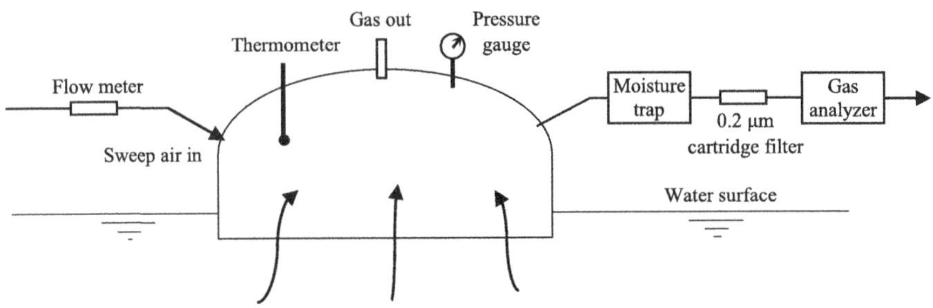

Figure 4.9 Schematic of static flux chamber with sweep gas.

The chamber has a hemispherical design and is a sealed chamber made of plastics, whose set-up is very similar to the SEIFC chamber. The chamber is held to a floating device and is attached to the tank skirts for position during the testing (Figure 4.10). The sweep gas is introduced to the chamber at a flow rate of 3–10 L/min. After the headspace air is exchanged at least five times, the measurement of gas concentrations in the exhaust gas commences. The emission flux of gas *i* from the covered surface is calculated as:

$$F_{E,i} = \rho_i \times c_i \times Q_{\text{sweepgas}} / A \times 10^{-6} \times 24 \times 3600 \quad (4.11)$$

where

Q_{sweepgas} = sweep gas flow rate (m^3/s);
C_i = concentration of gas *i* (CH_4, CO_2 or N_2O) in the exhaust gas (ppmv); and
A = surface area enclosed by the chamber (m^2).

The advantage of this technique is that it is relatively simple but it is labor intensive and time consuming compared with the previous static chamber method. The sweep gas flow rate is required to cause sufficient mixing but not be so strong that it causes turbulence to the quiescence of the water surface.

4.2.3.2 Surface emission isolation flux chamber (SEIFC)

USEPA also recommended the SEIFC method to measure the off-gas flow rate. The only difference from the SEIFC protocol for the aerated tank is to provide a sweep gas in addition to the tracer gas to measure the off-gas flow rate. Because of the low off-gas flow rate, the mixing of the off-gas by using a sweep gas provides more accurate results. A sweeping gas, which is air or carrier gas, in addition with helium tracer gas (H_e), is introduced to the flux chamber. The procedure is similar to the SEIFC protocol for aerated tanks. From the concentrations of helium gas in the exhaust gas measured, the emission rate of GHGs from the covered areas will be calculated using Eq. 4.12 according to the helium gas' mass balance as illustrated in Figure 4.11.

Figure 4.10 Flux chamber used by French scientist (Foley *et al.* 2011).

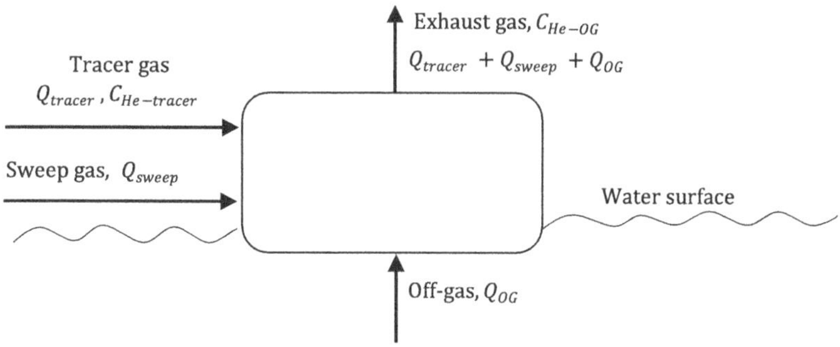

Figure 4.11 Mass balance for the SEIFC when testing the non-aeration tanks.

$$Q_{\text{tracer}} C_{\text{He-tracer}} = (Q_{\text{tracer}} + Q_{\text{sweep}} + Q_{OG}) C_{\text{He-OG}} \tag{4.12}$$

$$Q_{OG} = \frac{Q_{\text{tracer}} (C_{\text{He-tracer}} - C_{\text{He-OG}})}{C_{\text{He-OG}}} - Q_{\text{sweep}} \tag{4.13}$$

where

Q_{sweep} = sweep gas flow rate (m^3/s).

The sweep gas flow rate needs to be large enough to cause sufficient mixing in the chamber but not so strong that it causes perturbation of the quiescent water surface. A sweep gas flow rate between 5 and 13 L/min is appropriate (Eklund, 1992).

4.3 QUANTIFICATION OF N_2O EMISSION THROUGH MEASURING DISSOLVED N_2O CONCENTRATION IN THE LIQUID PHASE

4.3.1 N_2O emission and generation

If the soluble N_2O concentration in the liquid phase is known, N_2O emission to the atmosphere through diffusion and air stripping across the water-air interface can be estimated. According to Eq. 2.12 in Chapter 2, the diffusion and air striping (including diffusion) rates of N_2O in wastewater can be described using linear equations with respect to the dissolved N_2O concentration (Li, 2010):

$$r_e = -K_{La} C_{N_2O} \tag{4.14}$$

where

r_e = N_2O emission rate (kg/m^3/day);

C_{N_2O} = soluble N_2O concentration (kg/m^3); and

K_{La} = mass transfer coefficient (day^{-1}). A negative sign of r_e means that N_2O is emitted from the water phase to the atmosphere.

The accumulation rate, r_c, of soluble N_2O in the water tank, at time t, is calculated using Eq. 4.15:

$$r_c = \frac{dC_{N_2O}}{dt} \tag{4.15}$$

where, r_c is the N_2O accumulation rate in the liquid phase (kg/m^3/day). A positive r_c indicates that N_2O is accumulated in the water tank while a negative r_c indicates that N_2O concentrations decline in the water tank.

According to the mass balance for N_2O in the water tank, as described in Chapter 2, Eq. 2.4, an equation can be built:

$$Vr_c = Q_I C_{N_2O,i} - Q_I C_{N_2O,e} + Vr_e + Vr_g \tag{4.16}$$

where
r_g = N_2O generation rate (kg/m³/day).

Hence, r_g is calculated as:

$$r_g = r_c - \frac{Q_I C_{N_2O,i} - Q_I C_{N_2O,e}}{V} - r_e \tag{4.17}$$

A positive r_g indicates that N_2O is generated in the water tank in lane, while a negative r_g indicates that N_2O is reduced in the water tank due to biochemical reactions.

The specific N_2O generation rate can be calculated:

$$\mu_g = \frac{r_g}{MLVSS} \tag{4.18}$$

where
μ_g = specific N_2O generation rate (kg/kg VSS/day); and
$MLVSS$ = concentration of mixed liquor volatile suspended solids (kg VSS/m³).

From the N_2O emission rate and generation rate, the amount of N_2O emission and generation during the period of $t_1 - t_2$ are calculated as:

$$E = -V \int_{t_1}^{t_2} r_e dt \tag{4.19}$$

$$G = V \int_{t_1}^{t_2} r_g dt \tag{4.20}$$

where
E and G = amounts of N_2O emission and generation in the testing period (from t_1 to t_2), respectively (kg); and
V = effective tank volume (m³).

Compared with the off-gas measurement, the direct measurement of N_2O has a significant big advantage. Since N_2O has a relatively high solubility in wastewater, the off-gas measurement does not reflect when and where N_2O gas is generated in the whole wastewater treatment stream. For example, in the aeration tank, there might be a high N_2O emission due to the air stripping effects; however, the N_2O emitted might be generated in the preceding non-aerated tanks. In particular, since dissolved N_2O can be measured using developed sensors (Andersen *et al.* 2001; Kampschreur *et al.* 2007; Li, 2010), dissolved N_2O concentrations can be measured continuously online. For surface-aerated aeration tanks, measurement of GHGs emissions with the floating hood or flux chamber techniques is quite difficult to set up while measurement of soluble gases is easier to carry out.

4.3.2 Dissolved N_2O measurement with the air stripping technique

The dissolved N_2O concentration in the water samples can be stripped out to the headspace at an elevated temperature in acidic conditions (Foley *et al.* 2011). After measuring the N_2O concentration in the headspace, the dissolved N_2O concentration in the water phase can be calculated.

A certain volume of wastewater sample, grab or composite, is added to a serum bottle gently along the bottle side. Several drops of sulfuric acid are added to the bottle and then it is sealed airtight. The bottle is placed in a water bath and heated at 80°C for 1 h so that N_2O is released to the headspace completely. The gas samples in the headspace are taken with gas syringes and then analyzed with gas chromatography (Foley *et al.* 2011). The concentration of dissolved N_2O in the wastewater sample is calculated as:

$$VC_{N_2O} = V_{\text{headspace}} C_g \times \frac{PT_0}{P_0 T} \div 22.4 \times 44 \times 10^{-6} \tag{4.21}$$

$$C_{N_2O} = \frac{V_{\text{headspace}}}{V} \times C_g \times \frac{PT_0}{P_0 T} \div 22.4 \times 44 \times 10^{-6} \tag{4.22}$$

where

V = water sample volume (L);
$V_{\text{headspace}}$ = volume of the headspace in the bottle (L);
C_g = N_2O concentration in the headspace (ppmv);
P and T = pressure and temperature in the headspace (atm and K);
P_0 and T_0 = pressure and temperature at STP (1 atm and 273 K);
22.4 L/mol = molecular volume of gas at STP; and
44 g/mol = molecular mass of N_2O.

In order to measure C_g accurately, $V_{\text{headspace}}$ should not be much larger than V. The Rennes CEMAGREF research team used V of 100 ml and $V_{\text{headspace}}$ of 230 mL. The suitable range of the ratio $V_{\text{headspace}}$:V should be studied in future. This headspace method is simple. The problem is that only discrete testing can be carried out. During the sampling of wastewater from the wastewater tanks, it is very important to avoid turbulence since it would cause the loss of N_2O gases from the wastewater.

4.3.3 Dissolved N_2O measurement using N_2O sensor

A widely used sensor to measure dissolved N_2O is the one developed by Unisense (Denmark), which is the miniature Clark-type sensor constructed of glass (Figure 4.12) based on the diffusion of N_2O through a membrane in the sensor tip. In the sensor tip, the diffused N_2O is reduced at a cathode, generating an electric current which is converted to a signal. The signal is recordable and is able to represent the concentrations of soluble N_2O.

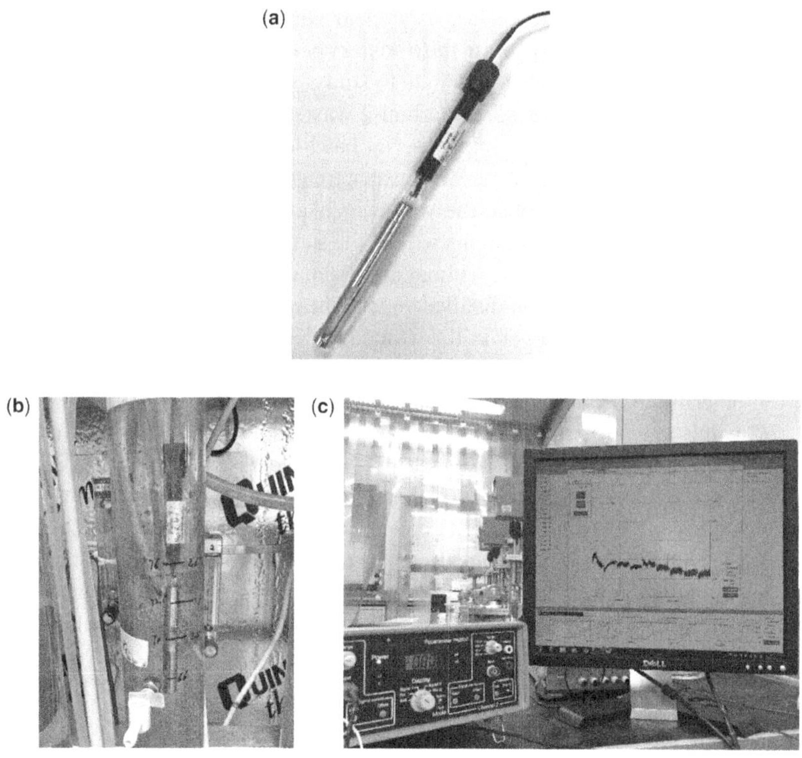

Figure 4.12 N_2O sensor system for testing soluble N_2O provided by Unisense. (a) N_2O sensor; (b) set-up of N_2O sensor in a lab-scale sequencing batch reactor; (c) signals display on the computer with a picoammeter on the left.

The N_2O sensor is able to accurately measure N_2O concentrations in the range 0–30 mg/L with a detection limit of 0.1 μM. Before the measurement, the sensor must be calibrated using two points. The zero reading is obtained by immersing the N_2O sensor in N_2O-free water. The second point reading is obtained by allowing the N_2O sensor to take a signal from an N_2O solution with known N_2O concentration, for example, 4.4 mg/L, which is diluted from the saturated N_2O solution.

Then the N_2O sensor is installed in selected locations in the water tank and records the soluble N_2O concentrations with the connection to a picoammeter and a computer.

In order to get a full picture of N_2O emissions and generations, the testing consists of several steps:

Step 1: Determine K_{La} by clear water test.

Li (2010), Zhang (2012) and Pan *et al.* (2014) used this technique to measure dissolved N_2O emissions from lab-scale intermittently aerated sequencing batch reactors (IASBRs). They used clear water, rather than wastewater, to calibrate the sensor and obtain the transfer coefficient K_{La}. This simplification allowed them to obtain enough data to study N_2O emission characteristics in IASBR systems treating slaughterhouse wastewater.

In the clear water test methodology, serial batch experiments are carried out to simulate the operation of the wastewater treatment tanks using tap water. A reactor is set up to simulate the wastewater tank that is to be studied. In the reactor, the operating conditions will be the same as those in the tank but there is no biomass. The reactor, without any biomass, is filled with a known N_2O concentration, made from distilled water. The reactor is aerated with the same aeration superficial flux as in the aerated water treatment tank or laboratory-scale reactor (m^3 air/m^2 tank surface/second). Under non-aeration (only stirring; to determine K value via diffusion) and aeration (with stirring; to determine K value via air stripping) conditions, the dissolved N_2O concentration is measured and recorded *in situ*, and the N_2O decrease rate, which is equal to the emission rate because no N_2O generation occurs, is calculated.

Step 2: Measure the dissolved N_2O concentrations in the reactors. N_2O sensors are set up in the selected locations in the wastewater treatment tanks. At least three locations should be chosen to represent the N_2O emissions for the targeted wastewater treatment tank. Since sensors can measure soluble N_2O continuously, there is no problem in obtaining N_2O data for 24 h.

Step 3: Use equations introduced in Section 4.3.1 to calculate N_2O emission rate, generation rate, and the amount of N_2O emission and generation in the testing period from the wastewater tank.

Case study: Using this technology, Zhang *et al.* (2012) studied N_2O emissions from three laboratory-scale aerobic granular sludge sequencing batch reactors (SBRs). The three identical SBRs (SBR1, SBR2 and SBR3) were made from transparent Plexiglas, each having an effective volume of 1.5 litres (internal diameter = 5 cm and height = 100 cm). They were operated under the same conditions except for different air supply rates. Air was supplied using aquarium air pumps through air diffusers installed at the bottom of the reactors, and the air flow rates to SBR1, SBR2 and SBR3 were 0.2, 0.6 and 1.0 L/min, respectively. The ambient temperature was 14 ± 4°C. The SBRs were operated in 4-hour cycles with a hydraulic retention time of 8 hours and a 50% volume exchange ratio each time. An SBR operation cycle consisted of 5 minutes fill, 220 minutes aeration, 5 minutes settlement and 10 minutes withdraw. The three reactors treated synthetic wastewater, which was made from tap water by adding glucose (COD = 700 mg/L), $(NH_4)_2SO_4$ (NH_4^+-N = 148 mg/L), 100 mg/L KH_2PO_4, 500 mg/L K_2HPO_4, 100 mg/L NaCl, 200 mg/l $MgCl_2$, 20 mg/L $FeCl_2$ and 20 mg/L $CaCl_2$. The reactors were operated as aerobic granular sludge bioreactors and the biomass concentrations in the SBRs were 18.9 ± 0.8 g suspended solids (SS)/L, and the mean particle diameter was 710 µm.

First, K_{La} was determined. Clear water test experiments mimicked the operation of the three reactors. An N_2O solution made of distilled water with an N_2O concentration of 6.2 mg/L was gently added into the three reactors without any biomass. The reactors were aerated at aeration rates of 0.2, 0.6 and 1.0 L/min, respectively. The N_2O concentrations were recorded using the N_2O sensor. The N_2O emission rates were linearly dependent on the N_2O concentration (Figure 4.13):

$r_e = -0.106C_{N_2O}$ at the aeration rate of 0.2 L/min
$r_e = -0.131C_{N_2O}$ at the aeration rate of 0.6 L/min
$r_e = -0.150C_{N_2O}$ at the aeration rate of 1.0 L/min

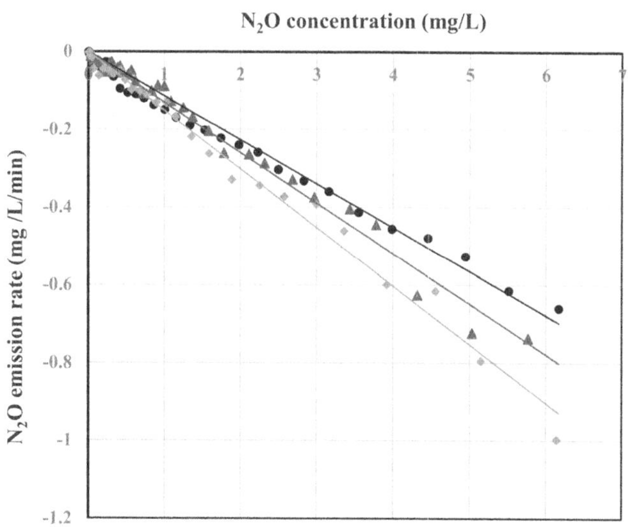

Figure 4.13 Linear dependence of N_2O emission rates on the dissolved N_2O concentration in the clear water test (●: 0.2 L/min, ▲: 0.6 L/min, ◆: 1.0 L/min) (Zhang *et al.* 2012).

Thus, the measured K_{La} was 0.106/min, 0.131/min and 0.150/min at the three aeration rates, respectively.

During the stable operation periods of the three aerobic granular sludge SBRs, real-time N_2O concentrations were monitored. The typical cycle profiles of N_2O concentrations are presented in Figure 4.14. The dissolved N_2O concentration increased immediately after wastewater was fed.

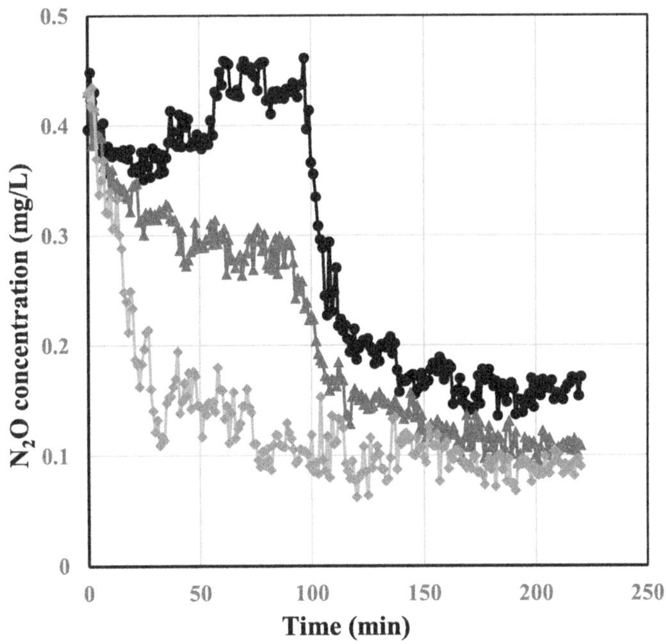

Figure 4.14 Profiles of N_2O in a typical cycle (●: 0.2 L/min, ▲: 0.6 L/min, ♦: 1.0 L/min).

At 2–5 minutes after the aeration commenced, the N_2O concentration decreased and leveled off for 30–50 min, before it dropped in SBR1 and SBR2. Using Eqs. 4.13 and 4.14, N_2O emission and generation rates were calculated (Figures 4.14 and 4.15). In the aeration stage, in SBR1 and SBR2, the average specific N_2O generation rate calculated was 2.1 μg N_2O/(g SS · min) and 2.2 μg N_2O/(g SS · min), respectively. In SBR3, dissolved oxygen (DO) concentrations increased sharply after the aeration commenced and were saturated in the liquid phase after 10–20 minutes' aeration. The N_2O concentration dropped after the aeration commenced. The average specific N_2O generation rate decreased to 1.1 μg N_2O/(g SS · min). In SBR1, SBR2 and SBR3, the effluent N_2O concentration was 0.17 mg/L, 0.11 mg/L and 0.09 mg/L, equal to 0.13%, 0.08% and 0.06% of the total nitrogen in the effluent, respectively. In accordance with Eq. 4.18, the N_2O emission amounts in a typical cycle were calculated. The proportions of N_2O emissions to the influent nitrogen loading rate (NLR) in SBR1, SRB2 and SBR3 were 8.2%, 6.1% and 3.8%, respectively.

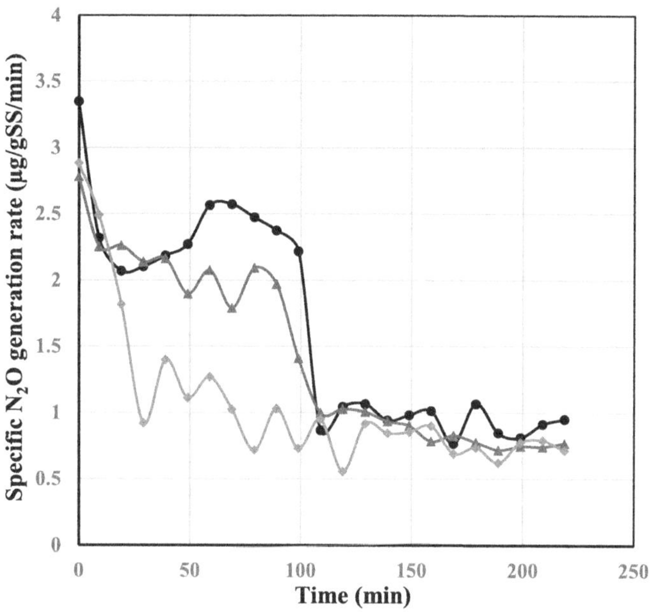

Figure 4.15 Specific N_2O generation rate in a typical cycle (●: 0.2 L/min, ▲: 0.6 L/min, ◆: 1.0 L/min).

4.3.4 Estimation of K_{La} combining dissolved N_2O concentration measurement and off-gas measurement

In the method discussed in Section 4.3.3, it is assumed that K_{La} values measured through the clear water tests are equal to those of wastewater treatment systems, based on the finding of Bielefeldt and Stensel (1999) (described in Chapter 2). Another assumption is that the saturation concentration of N_2O in the clear water is the same as that in the wastewater. Both assumptions are not accurate to some extent. The saturation concentration of N_2O concentration in the wastewater would be reduced to 95% in comparison with clear water (Tchobanoglous *et al.* 2003). The K_{La} values measured in wastewater treatment tanks are also found to be lower than those values obtained with clear water tests (Foley *et al.* 2011). A more accurate K_{La} can be measured by combining the off-gas measurement with the floating hood or flux chamber techniques and the dissolved N_2O gas concentration method (Ye *et al.* 2014). When the off-gas measurement is carried out, the N_2O sensor is installed in the location beside the chamber.

If the off-gas measurement shows that N_2O emission flux is F_E (kg N_2O/m^2 tank surface/day), then

$$F_E = -H\frac{\int_{t_1}^{t_2} K_{La}(C - C^*)dt}{t_2 - t_1} \tag{4.23}$$

$$K_{La} = -\frac{F_E \times (t_2 - t_1)}{H\int_{t_1}^{t_2}(C - C^*)dt} \tag{4.24}$$

where

H = depth of the wastewater treatment tank;
t_1 and t_2 = dissolved N_2O testing start time and end time, respectively; and
C^* = saturated N_2O concentration in the wastewater.

If N_2O emission and dissolved N_2O concentration are steady in the testing period, then the equation can be simplified as:

$$K_{La} = -\frac{F_E}{H \times (C - C^*)} \tag{4.25}$$

After K_{La} is known, an N_2O sensor can be applied to the field study as described in Section 4.3.1 and N_2O can be monitored continuously.

4.4 MEASUREMENT OF DISSOLVED METHANE CONCENTRATION

There are two methods for measuring dissolved methane concentrations in the water phase: headspace method and salting-out method. Other methods are modified based on these two methods.

4.4.1 Headspace method

Wastewater is gently added into a serum bottle with a capacity of V (L) along the side of the bottle so as to avoid turbulence (Figure 4.3) until the bottle is full. Then several drops of preservative (20 mM mercury (II) chloride or 1:1 sulfuric acid) are added to the bottle to stop biochemical activities that may generate or consume CH_4. Then the bottle is sealed. Helium gas is introduced to the bottle to push 10% of the water out of the bottle and create a headspace of 10% V (Kampbell & Vandegrift, 1998). The gas-liquid mix is then vigorously shaken for 10 minutes so that equilibrium between the gases in the headspace and dissolved gases in the water phase is reached. The gas in the headspace is sampled with a syringe and then the methane concentration in the gas is measured by gas chromatography. The dissolved CH_4 concentration is then calculated based on Henry's Law.

At the equilibrium, the concentration of dissolved CH_4 in the water phase, g/L, is equal to:

$$C_L = \frac{C_g}{H} \times 55.6 \times 16 \tag{4.26}$$

where

C_g = methane concentration in the gas (V/V);
H = Henry's law constant for gas;
55.6 is moles of water molecules in 1 litre of water; and
16 is the molecular mass of CH_4 (16 g/mole).

A mass balance can be built based on the dissolved methane before and after the equilibrium is reached, as illustrated with Figure 4.16:

$$C \times 0.9V = C_L \times 0.9V + \frac{C_g \times 0.1V \times P \times T_0}{P_0 \times T \times 22.4} \times 16 \tag{4.27}$$

where

P_0 and T_0 = pressure and temperature at STP (1 atm and 273 K) respectively;
P and T = atmospheric pressure and ambient temperature during the test, respectively; and
22.4 is the molar volume of a gas at STP (22.4 L/mole).

Hence,

$$C = 889.6\frac{C_g}{H} + 0.071\frac{C_g VPT_0}{P_0 T} \tag{4.28}$$

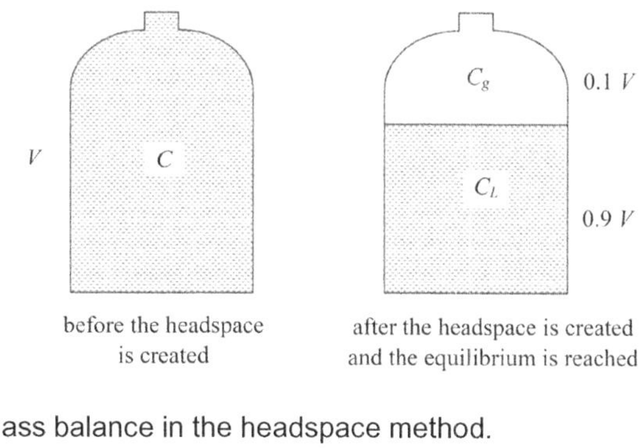

Figure 4.16 Mass balance in the headspace method.

4.4.2 Salting-out method

In the sampling method as illustrated in Figure 4.17, a serum bottle, whose capacity is V (L), is added with solid NaCl and KOH (which will be equal to 400 g/L NaCl and 60 g/L KOH concentration when the bottle is full of water). Then wastewater with a sample size of V_1 L ($V_1 < V$) is added into the serum bottle gently along the side of the bottle so as to avoid turbulence or the wastewater is added with a syringe with a catheter tip and a silicone tube (Daelman, 2014). The water surface is marked on the serum bottle. Then the bottle is sealed. The gas-liquid mix is then vigorously shaken or kept in a boiling water bath for 2 h so that equilibrium between the headspace and the water space is reached (Gal'chenko *et al.* 2004). The gas composition in the headspace is analyzed by gas chromatography. After the gas sample is taken, the liquid in the serum bottle is replaced with the clear water until the mark is reached. The volume of the clear water added is measured, V_3, so the headspace volume V_2 is equal to $V - V_3$. NaCl is over saturated so the dissolved CH_4 is salted out. It will also inhibit the biochemical reactions that generate or consume CH_4. The dissolved CH_4 concentration is calculated as:

$$C \times V_1 = V_2 \times C_g \times \frac{PT_0}{P_0T} \div 22.4 \times 16 \tag{4.29}$$

$$C = \frac{V_2C_g}{V_1} \times \frac{PT_0}{P_0T} \div 22.4 \times 16 \tag{4.30}$$

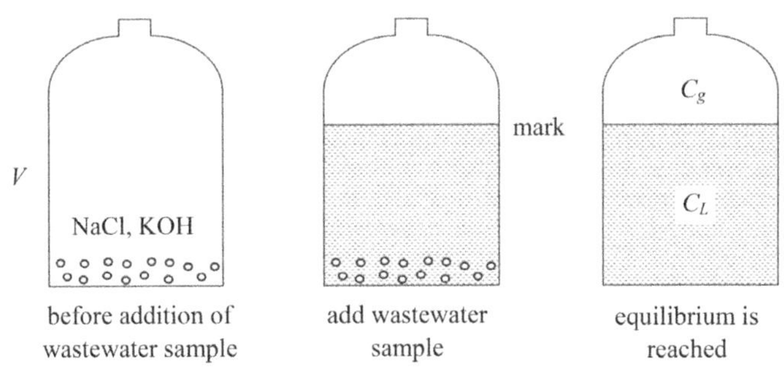

Figure 4.17 Schematic of the salting-out method.

In the experiment, KOH is added with the purpose to absorb soluble CO_2 in the water sample. If KOH is not added, then release of soluble CO_2 from the water phase to the air causes obvious pressure increase in the headspace so it is necessary to measure the headspace volume expansion due to pressure built up in the bottle (Daelman, 2014).

4.5 SUMMARY

Measurement of direct GHG emissions can provide policymakers with the amount of GHG emissions from wastewater treatment facilities, and determine GHG emission factors for national GHG inventory estimations. It also provides the plant operators with information on how the WWTPs are performing. For instance, when the aerated nitrifying tanks receive high loading shocks or air blowers fail leading to insufficient oxygen supply, N_2O emissions will significantly increase. Therefore, the unexpectedly increased N_2O emissions will tell the operators that the nitrifying tank is not successful.

It is recommended that when measuring the direct GHG emissions from WWTPs, liquid samples are also taken from the targeted WWTP tanks for measuring COD, BOD_5, NH_4^+, NO_2^-, NO_3^-, alkalinity, pH and DO, etc. These parameters should be linked to the GHG emissions data so as to understand the GHG generation mechanisms. If direct GHG emission is measured with the floating hood or SEIFC chamber techniques, liquid samples should be taken beside the locations where the floating hood and static chamber are installed. If direct GHG emission is monitored by measuring dissolved gases, then liquid samples should be taken for analyzing these parameters. In addition, DO, pH and oxidation-reduction potential (ORP) probes should be set up in these locations to obtain DO, pH and ORP data.

Before off-gas and liquid sampling campaigns commence, information regarding the WWTP operation should be collected. This information should include: historical data of wastewater flow rates, water characteristics, industrial and commercial water flows, sewer system, local temperature, etc. A complete GHG emission campaign contents are summarized in Figure 4.18.

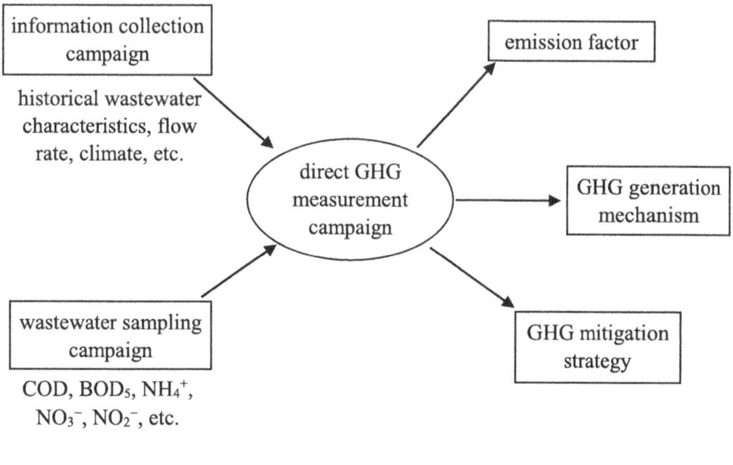

Figure 4.18 Complete GHG emission campaigns.

4.6 REFERENCES

Aboobakar A., Jones M., Vale P., Cartmell E. and Dotro G. (2014). Methane emissions from aerated zones in a full-scale nitrifying activated sludge treatment plant. *Water, Air & Soil Pollution*, **225**(1), 1814.

Andersen K., Kjær T. and Revsbech N. P. (2001). An oxygen insensitive microsensor for nitrous oxide. *Sensors and Actuators B: Chemical*, **81**(1), 42–48.

Bielefeldt A. R. and Stensel H. D. (1999). Treating VOC-contaminated gases in activated sludge: mechanistic model to evaluate design and performance. *Advances in Environmental Science and Technology*, **33**(18), 3234–3240.

Carney K. N. (2012). Treatment of Piggery Wastewater Using Woodchip Biofilters. PhD thesis, National University of Ireland, Galway, Ireland.

Chandran K., Volcke E. I. P. and Van Loosdrecht M. C. M. (2016). Experimental Methods in Wastewater Treatment: Off-Gas Emission Tests. IWA Publishing, London.

Czepiel P. M., Crill P. M. and Harriss R. C. (1993). Methane emissions from municipal wastewater treatment processes. *Advances in Environmental Science and Technology*, **27**(12), 2472–2477.

Daelman M. (2014). Emissions of Methane and Nitrous Oxide from Full-Scale Municipal Wastewater Treatment Plants. PhD thesis. TU Delft.

Eklund B. (1992). Practical guidance for flux chamber measurements of fugitive volatile organic emission rates. *Journal of the Air and Waste Management Association*, **42**(12), 1583–1591.

Foley J., Yuan Z., Keller J., Senante E., Chandran K., Willis J., Shah A., van Loosdrecht M. and van Voorthuizen E. (2011). N_2O and CH_4 Emission from Wastewater Collection and Treatment Systems: Technical Report. Global Water Research Coalition, London, United Kingdom.

Gal'chenko V., Lein A. Y. and Ivanov M. (2004). Methane content in the bottom sediments and water column of the Black Sea. *Microbiology*, **73**(2), 211–223.

Hutchinson G. and Mosier A. (1981). Improved soil cover method for field measurement of nitrous oxide fluxes. *Soil Science Society of America Journal*, **45**(2), 311–316.

Kampbell D. H. and Vandegrift S. A. (1998). Analysis of dissolved methane, ethane, and ethylene in ground water by a standard gas chromatographic technique. *Journal of Chromatographic Science*, **36**(5), 253–256.

Kampschreur M. J., Tan N. C., Kleerebezem R., Picioreanu C., Jetten M. S. and Loosdrecht M. C. V. (2007). Effect of dynamic process conditions on nitrogen oxides emission from a nitrifying culture. *Advances in Environmental Science and Technology*, **42**(2), 429–435.

Li J. (2010). Nutrient Removal from Slaughterhouse Wastewater and Oxygen-Limited Partial Nitrification Process in Sequencing Batch Reactors. PhD thesis, National University of Ireland, Galway, Ireland.

Mosier A., Duxbury J., Freney J., Heinemeyer O. and Minami K. (1996). Nitrous oxide emissions from agricultural fields: assessment, measurement and mitigation. *Plant and Soil*, **181**(1), 95–108.

Pan M., Wen X., Wu G., Zhang M. and Zhan X. (2014). Characteristics of nitrous oxide (N_2O) emission from intermittently aerated sequencing batch reactors (IASBRs) treating slaughterhouse wastewater at low temperature. *Biochemical Engineering Journal*, **86**, 62–68.

Pan Y., van Den Akker B., Ye L., Ni B.-J., Watts S., Reid K. and Yuan Z. (2016). Unravelling the spatial variation of nitrous oxide emissions from a step-feed plug-flow full scale wastewater treatment plant. *Scientific Reports*, **6**, 20792.

Tchobanoglous G., Burton F. L., Stensel H. D. and Metcalf and Eddy Inc. (2003). *Wastewater Engineering: Treatment and Reuse*. McGraw-Hill Education, New York.

Ye L., Ni B.-J., Law Y., Byers C. and Yuan Z. (2014). A novel methodology to quantify nitrous oxide emissions from full-scale wastewater treatment systems with surface aerators. *Water Research*, **48**, 257–268.

Zhang M. (2012). Treatment of Pig Manure Digestate Liquid and Study of Nitrous Oxide Emissions from Bioreactors. PhD thesis, National University of Ireland, Galway, Ireland.

Zhang M., Lawlor P. G., Li J. and Zhan X. (2012). Characteristics of nitrous oxide (N_2O) emissions from intermittently-aerated sequencing batch reactors treating the separated liquid fraction of anaerobically digested pig manure. *Water, Air & Soil Pollution*, **223**(5), 1973–1981.

Chapter 5

Methane emission and mitigation in municipal wastewater treatment plants

Zhenhu Hu and Rui Tang

5.1 INTRODUCTION

Direct CH_4 emission is higher than N_2O emission in most wastewater treatment facilities, in terms of CO_2 equivalent. Up to now, the research investigating CH_4 emission from full-scale municipal wastewater treatment plants (WWTPs) has been conducted in very few countries. Through thorough understanding of the CH_4 emission mechanisms and pathways from full-scale WWTPs, mitigation strategies can be developed. In this chapter, the generation and emission of methane from municipal WWTPs, and the mitigation of methane emission, are discussed.

5.2 METHANE EMISSION IN WWTPs

To the best of our knowledge, the research reporting the total methane emissions from full-scale municipal WWTPs is very limited. Czepiel *et al.* (1993) investigated the methane emissions of a small WWTP in Durham, NH, USA, whose capacity was 4×10^3 m^3/day with a 12,500 population equivalent (PE). Wang *et al.* (2011a) studied the methane emissions of a large WWTP (1,500,000 PE) using an A^2O process in Jinan, China with a treatment capacity of 3×10^5 m^3/day. Liu *et al.* (2014a) reported the methane emissions of two WWTPs in Beijing, China with the treatment capacity of 5×10^5 m^3/day (1,200,000 PE; A^2O process) and 8×10^4 m^3/day (231,000 PE; sequencing batch reactor (SBR) process), respectively. In these investigations, grab sampling techniques were used to monitor the direct off-gas emission in the WWTPs. Daelman *et al.* (2012a) studied the methane emissions at Kralingseveer WWTP in Capelle aan den IJssel, NH with a capacity

Xinmin Zhan, Zhenhu Hu, Guangxue Wu
doi: 10.2166/9781780406312_079

of 360,000 PE. This plant has an anaerobic digestion facility for surplus sludge so the methane emission from the sludge digester was also included in their study. Consequently, the overall methane emission was far higher than the other studies mentioned above. The grab sampling technique was used for the determination of dissolved methane in the liquid. The off-gas was detected using on-line detection because the whole WWTP was housed except for the secondary settling tank.

5.2.1 Total methane emission from municipal wastewater treatment plants

Total methane emissions from municipal WWTPs can be expressed in terms of kg CH_4/day, g CH_4/m^3 influent wastewater, kg CH_4/kg incoming chemical oxygen demand (COD), or g CH_4/(capita · year). According to the reports published in previous literature, the values reported vary significantly, depending on the wastewater treatment processes used, the incoming wastewater's characteristics, temperature, and operation of the WWTPs. Daelman *et al.* (2012a) measured the methane emission from the entire Kralingseveer WWTP from 14 October 2010 to 28 September 2011. The daily average methane emission was 302 kg CH_4/day with a standard deviation of 83 kg CH_4/day. The CH_4 emission factor was 306 g CH_4/(capita · year), or 3.44 g CH_4/m^3 influent. A total of 1.13% of the incoming COD was converted to methane. In Czeplel *et al.*'s research (1993), which is mentioned above, the emission factors were 39 g CH_4/(capita · year), 0.140 m^3 CH_4/m^3 influent, or 0.16% of the incoming COD. In the investigation by Wang *et al.* (2011a), the corresponding emission factors were 11.0 g CH_4/(capita · year), 0.160 m^3 CH_4/m^3 influent, or 0.08% of the incoming COD. In the investigation of Liu *et al.* (2014), the emission factors in the A^2O process were 0.182 g CH_4/m^3 influent and 24.8 g CH_4/(capita · year), while the emission factors for the SBR process were 0.457 g CH_4/m^3 influent and 36.5 g CH_4/(capita · year). The methane emission factors at Kralingseveer wastewater treatment plant are significantly high, which is due to the CH_4 loss to the atmosphere from the anaerobic digesters.

5.2.2 A/O process WWTPs

Nitrifying activated sludge processes have been adopted for wastewater treatment from an early stage in the development of municipal WWTPs. They comprise anoxic and oxic stages, and are therefore referred to as A/O processes. In the A/O process WWTPs, the removal of phosphorus is very limited. Analysis of CH_4 emissions from A/O process WWTPs has been limited thus far to just a single study (Aboobakar *et al.* 2014) in the literature.

In the study of Aboobakar *et al.* (2014), the A/O process unit is a plug-flow lane with a capacity of 210,000 PE. The emitted gas samples were collected from the lane which was divided into eight zones based on sampling location, as shown in Figure 5.1. The emitted gases were collected by the floating hood technique.

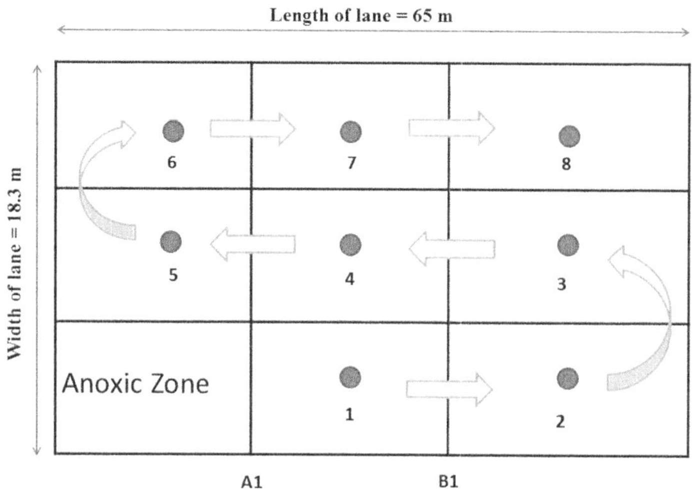

Figure 5.1 Schematic diagram of the plug-flow nitrifying lane reported by Aboobakar *et al.* (2014). The dots indicate sampling locations.

In the anoxic zone, the CH_4 emissions were considerably affected by the inflow rate from midnight to noon, but less affected from noon to midnight. In the oxic zone, CH_4 emissions had an irregular pattern, showing a weak correlation with the dissolved oxygen (DO) concentration. The inlet of the lane had high organic and ammonia loads and low DO concentrations, but had high CH_4 emissions. The CH_4 emissions were obviously correlated with the air flow rates (Figure 5.2). High air flow rates resulted in higher CH_4 emissions. The upstream of the lane had a high air flow rate and CH_4 emission rates were 8–16 g/h. The highest emissions occurred two thirds of the way down the aerated tank, where the CH_4 emission rates reached 20–24 g/h. The CH_4 emission occurred in the whole lane, and the highest emission was monitored in the later part of the aerated zone. Because most of the dissolved influent CH_4, which was generated in the sewer network, had been stripped out in the inlet of the lane, the emitted CH_4 in the later part should be generated in the lane. The generation of CH_4 in the aerobic zone can be explained by the structure of activated sludge flocs. The activated sludge flocs are divided into two zones: aerobic outermost zone and anaerobic innermost zone; the aerobic microorganisms exist in the outermost zone and the anaerobic microorganisms exist in the innermost zone. In Aboobakar *et al.*'s (2014) study, the dissolved oxygen level in the aeration lane was controlled at 1.5 mg O_2/L through employing blowers in duty/assist/standby mode. Therefore, methanogens were able to exist in the innermost zone, and continually generated CH_4 in the lane. In this A/O process, the CH_4 emission factor, in terms of a percentage of influent COD, ranged from 0.04 to 0.10% of influent COD, with 0.07% on average.

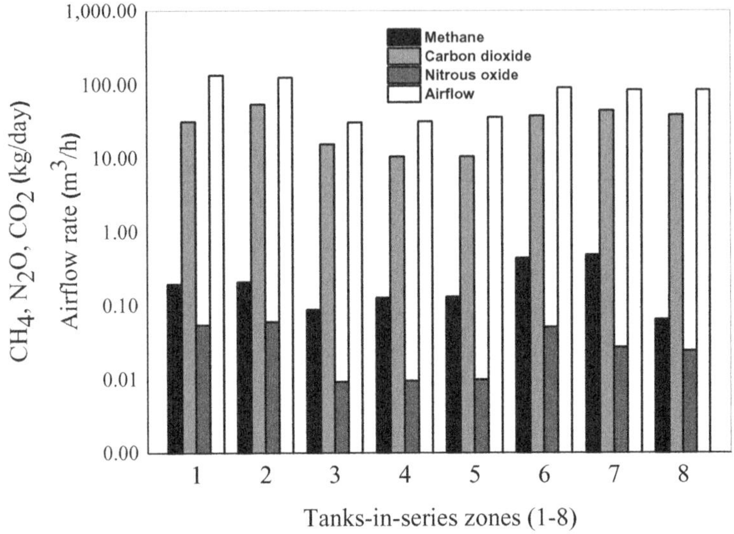

Figure 5.2 Spatial profile of average CH_4 emissions and air flow rates obtained over 8-week trial for the aerated lane (Aboobakar *et al.* 2014). The zone numbers are the sample collection points corresponding to Figure 5.1.

5.2.3 Sequencing batch reactor process WWTPs

The SBR process-based WWTP is one of the most popular secondary wastewater treatment processes. It is widely used in small and medium-size WWTPs, and accounts for 12% of built WWTPs in China (Liu *et al.* 2014). Liu *et al.* (2014) investigated the methane emission and conversion in an SBR process WWTP in Beijing, China with a treatment capacity of 231,000 PE (8×10^4 m^3/day). Figure 5.3 shows the process scheme of the WWTP. The bag sampling technique was used to measure the CH_4 fluxes from the aerated surfaces; the enclosed static chamber technique was used to collect the methane fluxes for the non-aerated wastewater surface; and the headspace method was used to measure dissolved methane in wastewater with the addition of mercury (II) chloride (Liu *et al.* 2014). The methane emission in the WWTP was monitored for one year. Methane emissions from individual tanks and the SBR phases are listed in Table 5.1. In this WWTP, more than 99.5% of the methane mission was detected in the feeding and aeration phases. However, the methane emissions fluctuated greatly in individual units during the sampling campaign period.

A relatively high concentration of dissolved CH_4 (0.66 mg/L) was detected in the swirl grit tank, which was generated in the sewer network before entering the WWTP (Liu *et al.* 2016). Therefore, the CH_4 emission in the swirl grit tank, where wastewater was intensively stirred so as to remove small particles, was correlated with CH_4 generation in the sewer network (Liu *et al.* 2015). The CH_4 flux in the swirl grit tank was relatively high, up to 4.8 g/($m^2 \cdot$ day). Because land is very limited in many cities, the swirl grit tank design is widely adopted in many WWTPs to enhance fine particle solids removal, therefore the CH_4 emission from swirl grit tanks shall be paid considerable attention.

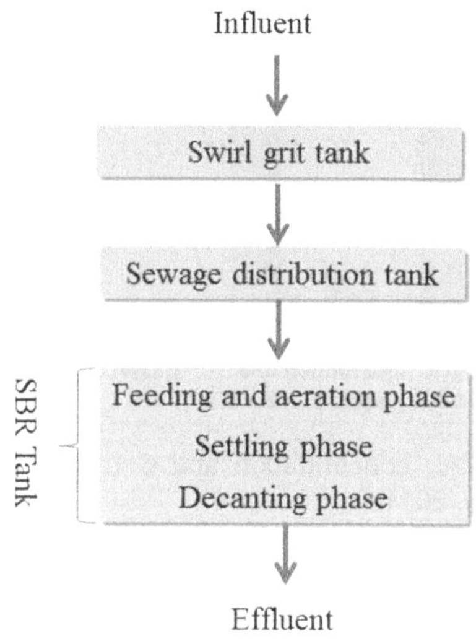

Figure 5.3 Process scheme of the SBR WWTP (Liu *et al.* 2014).

Table 5.1 Flow-normalized CH_4 emission from the units of an SBR process WWTP.

Processing Unit	Flow-normalized Emission (g/m^3 of wastewater)
Swirl grit tank	$6.19 \times 10^{-5} - 1.33 \times 10^{-3}$
Conditioning tank	$3.38 \times 10^{-5} - 1.27 \times 10^{-3}$
SBR feeding and aeration phase	$8.70 \times 10^{-2} - 1.59$
SBR settling phase	$1.01 \times 10^{-5} - 5.54 \times 10^{-4}$
SBR decanting phase	$1.06 \times 10^{-5} - 9.12 \times 10^{-4}$

Source: Liu *et al.* (2014).

In the SBR treatment process, 99.5% of the amount of CH_4 emissions occurred in the feeding and aeration phase of the SBR tank (Figure 5.4). The mechanical aeration stripped out dissolved CH_4 into the atmosphere. Therefore, the dissolved CH_4 concentration rapidly dropped during the feeding and aeration phases. The concentrations of dissolved CH_4 were very low in the settling and decanting phases, indicating CH_4 generation was low in the two phases, and the CH_4 emission fluxes in these phases were also low.

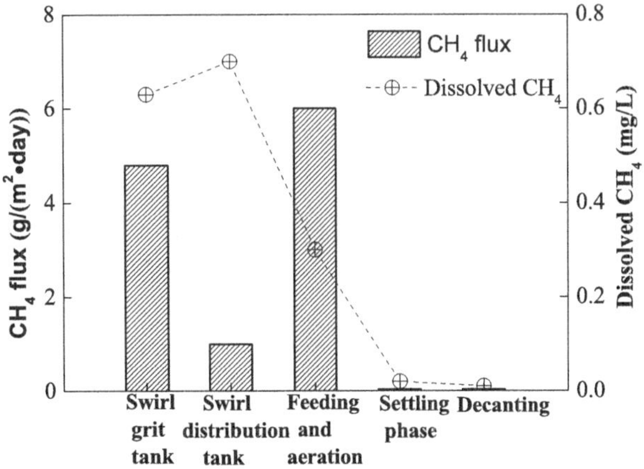

Figure 5.4 Dissolved CH_4 concentration and CH_4 flux in the SBR wastewater treatment plant (Liu *et al.* 2014).

5.2.4 A^2O process-based WWTPs

With the increasingly stringent effluent standards for wastewater treatment, biological nutrient removal processes have been widely used in full-scale municipal WWTPs. Many WWTPs have been upgraded to A/A/O processes, i.e., A^2O, to enhance nutrient removal, particularly nitrogen removal. This process has become the most popular process in large-scale WWTPs in China (Wang *et al.* 2011a). Compared with A/O process, an anaerobic unit is designed to enhance phosphorus removal, which may increase the generation of methane. Therefore, it is possible that more CH_4 would be emitted in the A^2O process.

Wang *et al.* (2011) assessed CH_4 emission of a full-scale municipal WWTP, based on the A^2O process, in Jinan, China. This plant had a treatment capacity of 1,500,000 PE (3×10^5 m^3/day) (the process scheme is shown in Figure 5.5). The bag sampling technique was used to measure the CH_4 fluxes for the aerated surfaces. The enclosed chamber technique was used to collect the methane fluxes for the non-aerated wastewater surface, and the headspace method was used to measure dissolved methane in wastewater with the addition of mercury (II) chloride.

The CH_4 emissions were monitored from March to June in the influent pump station, aerated grit chamber, anaerobic tanks, anoxic tanks, oxic tanks, secondary clarifiers, high density settler tanks, high efficiency fiber filter bed (for removing fine suspended solids (SS) before wastewater discharge), sludge thickening tanks, sludge screw conveyor, and sludge drying ground. The average CH_4 fluxes per unit surface area and the annual methane flux from each operation unit are shown in

Figure 5.6 and Table 5.2 respectively. The dominant CH_4 emission sources were (in descending order): anaerobic tanks > oxic tanks > aerated grit chambers > sludge thickening tanks. More than 60% of the total CH_4 emission in this WWTP came from the anaerobic and oxic tanks, because these two units had a large surface area and relatively higher emission flux per unit area (Wang *et al.* 2011a). The CH_4 emission in the pump station was 5.84 $g/m^2/day$ with the third highest methane influx per surface area among all the operation units in the WWTP. Because the surface area of the pump station is small, the total annual methane flux in the pump station was relatively low. The CH_4 emission factor for the whole plant was 0.155 g/m^3 influent (Wang *et al.* 2011a).

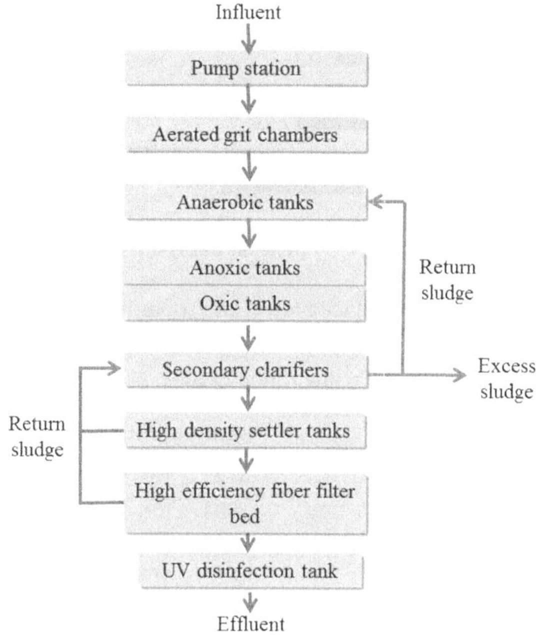

Figure 5.5 Process scheme of an A^2O process-based WWTP reported by Wang *et al.* (2011).

Liu *et al.* (2014a) also investigated the CH_4 emission in a full-scale A^2O process WWTP in Beijing, China, which had a treatment capacity of 1,200,000 PE (5 × 10^5 m^3/day). Figure 5.7 shows the process scheme of the A^2O WWTP. Measured direct CH_4 emissions from each operation unit are listed in Table 5.3, which are different from Wang *et al.* (2011). The mean dissolved CH_4 concentration in the influent wastewater entering the WWTP was 0.592 mg/L, indicating that a large quantity of CH_4 was generated in the sewer network. A high CH_4 oxidation rate of 32.5–89.5% was estimated according to CH_4 mass balance, which mainly occurred in the oxic tanks. The main CH_4 emission units in this WWTP were the primary clarifiers and oxic tanks, which were a little different from the research of Wang *et al.* (2011), in which the anaerobic tanks and oxic tanks were the main emission units. In the study of Liu *et al.* (2014), the WWTP has primary clarifiers with a large surface area, but in the study of Wang *et al.* (2011), there are no primary clarifiers.

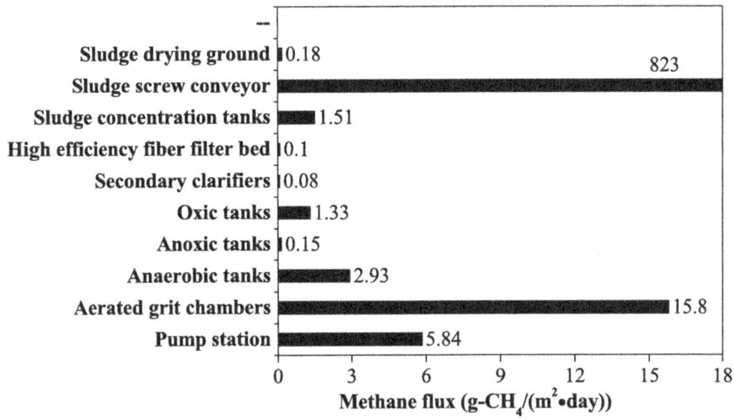

Figure 5.6 CH_4 fluxes per unit surface area in each operation unit (Wang *et al.* 2011a).

Table 5.2 Annual methane flux range of each processing unit.

Processing Unit	Annual Flux Range (kg/year)
Influent pump station	0.71×10^2–1.12×10^2
Aerated grit chamber	7.91×10^2–2.41×10^3
Anaerobic tanks	4.98×10^3–8.04×10^3
Anoxic tanks	2.53×10^2–4.14×10^2
Oxic tanks	3.84×10^3–9.44×10^3
Secondary clarifiers	2.80×10^2–5.14×10^2
High efficiency fiber filter bed	1.80×10–3.70×10
Sludge thickening tanks	1.13×10^3–2.00×10^3
Sludge screw conveyor	5.80×10–8.90×10
Sludge drying ground	1.00×10–2.10×10
Total	1.41×10^4–2.22×10^4

Source: Wang *et al.* (2011a).

The mean CH_4 emission factor was 0.182 g CH_4/m^3 of influent wastewater. The emission factor based on capita was 24.75 g CH_4/(capita · year), higher than the value of 11.3 g CH_4/(capita · year) in the study by Wang *et al.* (2011), but lower than that of 39 g/(capita · year) reported by Czepiel *et al.* (1993), which might be due to the difference of per capita water consumption and the dissolved CH_4 concentration in the influent wastewater (Liu *et al.* 2014). This clearly shows the high uncertainty in calculating CH_4 emission inventory when choosing emission factors.

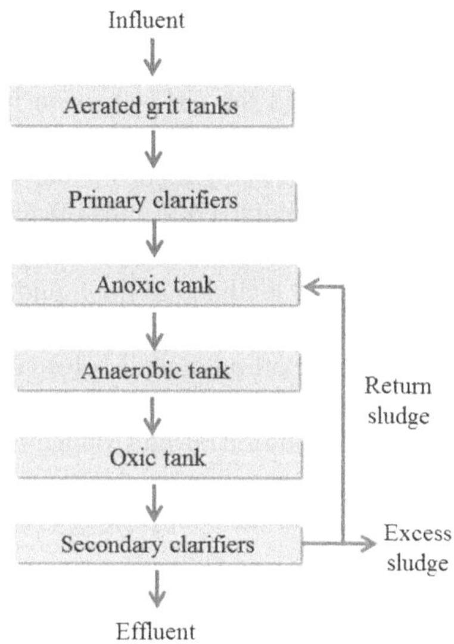

Figure 5.7 Process scheme of an A^2O WWTP (Liu *et al.* 2014).

Table 5.3 Flow-normalized CH_4 emission from the units of an A^2O process WWTP.

Processing Unit	Flow-normalized Emission (kg/year)
Aerated grit tanks	2.65×10^2–4.82×10^3
Primary clarifiers	1.13×10^3–2.34×10^4
Anaerobic tank	1.83×10^2–1.27×10^3
Anoxic tank	2.39×10^2–3.49×10^3
Oxic tank	5.86×10^3–6.79×10^4
Secondary clarifiers	3.30×10–3.89×10^2

Source: Liu *et al.* (2014).

5.2.5 Oxidation ditch process-based WWTPs

Aside from conventional activated sludge treatment, SBR, A/O, and A^2O wastewater treatment processes, oxidation ditches are also widely used for municipal wastewater treatment. For instance, this process is used in 29.2% of municipal WWTPs in China (Jin *et al.* 2014). Oxidation ditches are applied in a variety of sizes: small scale, medium scale, and large scale. The configurations of oxidation ditches include Orbal, Carousel, Triple channels oxidation ditch (T-OD), and Duel channels oxidation ditch (DE-OD) (Zhang *et al.* 2016). Although it is widely used, the methane emission from oxidation ditch-based WWTPs has only been reported in one paper by Daelman *et al.* (2012). In this research, the CH_4 emission from a Carousel oxidation ditch WWTP located in the municipality of Capelle ann den IJssel, in the Netherlands was measured.

Daelman *et al.* (2012) investigated the CH_4 emission in a plug-flow Carousel oxidation ditch WWTP. Figure 5.8 shows the WWTP process scheme. The WWTP has a treatment capacity of 360,000 PE. Except for the secondary clarifiers, all process units in the WWTP are covered to collect the off-gas for purification with a compost filter. The off-gas emissions in this WWTP were measured using an on-line infrared gas analyzer. The dissolved methane in the liquid phase was stripped out using the salting-out technique (Daelman *et al.* 2012). Therefore, the emission balance was established based on the off-gas and dissolved CH_4 in the liquid phases, through which the locations where CH_4 was generated or oxidized in the WWTP can be found. In the WWTP the produced surplus sludge is anaerobically digested for producing biogas, which is used to generate electricity on site. The electricity generated could provide about 60% of the plant's energy requirement.

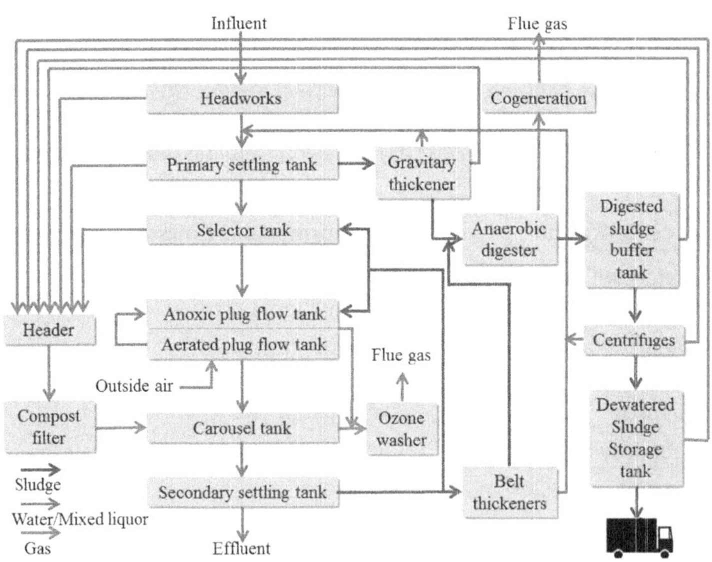

Figure 5.8 Process scheme of a Carousel oxidation ditch WWTP reported by Daelman *et al.* (2012).

The CH_4 emissions from the headworks, primary settling tank, selector tank, anoxic and aerated plug flow tanks, secondary settling tank, and anaerobic digester were measured. About 1% of the influent COD was emitted as CH_4, and the anaerobic digesters contributed to more than 80% of the total CH_4 emission. Table 5.4 lists the CH_4 emission from all process units. The CH_4 emission from the headworks was negligible, and most dissolved methane entered the primary clarifier. The CH_4 emission in the pump station depends on the type of pumps used. For centrifugal pumps, there is no contact between wastewater and air, but for screw conveyors, there is intense contact between wastewater and air, which causes significant air stripping of the dissolved CH_4 out of the wastewater. Therefore, mitigation of CH_4 emission in the pump station should be taken into account in the design of new WWTPs in the future. The daily CH_4 emission from this WWTP was 306 g CH_4/(capita · year) which also included CH_4 emission from the anaerobic digesters. The main CH_4 emissions occurred in the anaerobic digesters, Carousel oxidation ditch, digested sludge buffering tank, anoxic and aerated plug flow tanks, selector, and dewatered sludge storage tank. Collected CH_4, after compost filter treatment, was aerated into the Carousel oxidation ditch with an average flow rate of 288 kg/day, and the average CH_4 emitted from the Carousel oxidation ditch was 264 kg/day, thus the actual CH_4 emission in Carousel oxidation ditch was negative 24 kg/day. It is very interesting that most of the CH_4 from the influent of the wastewater was oxidized in the aeration sections, such as aerated plug flow tanks and Carousel oxidation ditch, not emitted to the atmosphere. This research provides the idea of reducing CH_4 mission by oxidizing CH_4 in aerated activated sludge tanks. In Daelman *et al.*'s (2012) study, the net CH_4 emission in the aeration sections was very low, which is quite different from other research results described in the preceding section.

Table 5.4 Methane emissions and balance from the various units (kg CH_4/day).

Operation Units	CH_4 Emission	Emission Balance*
Headworks	0 ± 0	-24 ± 96
Primary clarifier	24 ± 24	48 ± 96
Selector	48 ± 24	24 ± 96
Anoxic and aerated plug flow tanks	48 ± 24	-168 ± 72
Carousel oxidation ditch	264 ± 72	-24 ± 120
Secondary clarifier		24 ± 24
Belt thickeners	0 ± 0	0 ± 0
Gravitary thickener for primary sludge	24 ± 24	24 ± 24
Anaerobic digester	2928 ± 672	2928 ± 672
Digested sludge buffer tank	96 ± 24	96 ± 24
Centrifuges for sludge dewatering	24 ± 120	0 ± 0
Dewatered sludge storage tank	48 ± 24	48 ± 24
Compost filter	132 ± 24	0 ± 120
Electricity generation	48 ± 0	-2952 ± 672

*emission balance is the difference of methane flow between flow-in and flow-out in the unit.
Source: Daelman *et al.* (2012a).

5.3 PARAMETERS AFFECTING METHANE EMISSIONS IN WWTPs

Methane emission from WWTPs varies from plant to plant, and is affected by the features of each WWTP, i.e., the sewer network, wastewater characteristics (like C/N ratios), wastewater treatment processes, sludge management lines, aeration systems, and the use of anaerobic digestion facilities (Masuda *et al.* 2015). Emissions are also affected by the operating conditions, such as DO, temperature, and pH (Ren *et al.* 2015).

5.3.1 Effect of temperature and seasonal change

The wastewater temperature varies with local climate and season. It impacts the biochemical reaction kinetics in the processing units, which consequently affects the CH_4 generation and emission rates, and also impacts the diffusion rate of CH_4 from the wastewater phase to the air. High wastewater temperature results in high activities of methanogens, generating more methane than at low temperatures. Masuda *et al.* (2015) investigated the seasonal change of greenhouse gas (GHG) emission in a wastewater treatment plan with an A/O process. The CH_4 emission obviously varied with the change of season, as shown in Figure 5.9. The summer had the highest CH_4 emission rate, followed by fall, spring and winter. The wastewater had the highest temperature in summer while it had the lowest temperature in winter. There was an obvious correlation between wastewater temperature and CH_4 emission rates ($R^2 = 0.873$).

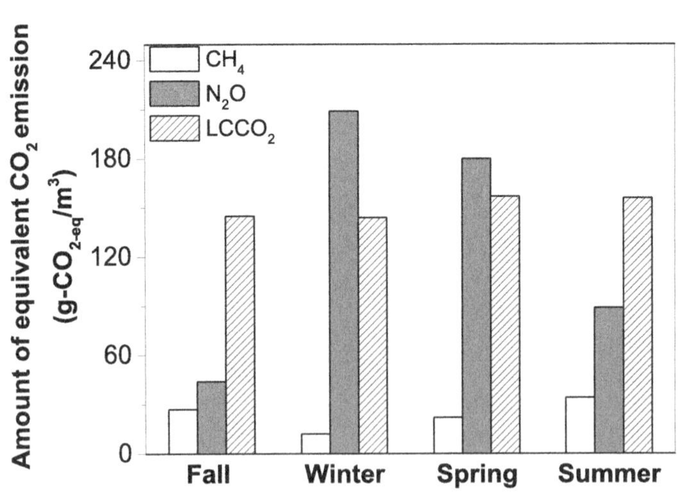

Figure 5.9 Seasonal variations in CH_4, N_2O, and CO_2 (Masuda *et al.* 2015). $LCCO_2$ is the amount of CO_2 emission estimated with life cycle assessment.

During Wang *et al.*'s (2011) research described in Section 5.2.4, the wastewater temperature rose from 12 to 24°C during the investigated period. They found that the impact of temperature on CH_4 emission depended on the operation units. There was obvious correlation between the wastewater temperature and CH_4 emission in the high density settler tanks ($R^2 = 0.67$), but no obvious correlation in other process units (Wang *et al.* 2011a). However, Aboobakar *et al.* (2014) didn't find a correlation between the temperature and CH_4 emissions, which was attributed to the short experimental duration and relatively stable wastewater temperature (ranging from 17.9 to 18.1°C) during the experimental period (Aboobakar *et al.* 2014).

5.3.2 Dissolved oxygen concentration

Because methanogens are strictly and obligately anaerobic microorganisms, the presence of oxygen inhibits methane generation in WWTPs. High DO inhibits methane generation while in order to maintain high DO levels in the tanks, high air flow rates have to be applied, which enhances the air stripping effects of CH_4 from the tanks to the air. Hence, the relationship between DO concentrations and CH_4 emission is difficult to conclude. Wang *et al.* (2011) reported the relationship between DO concentration and CH_4 emission in the Jinan WWTP. Intensive mechanical surface aeration resulted in high DO concentrations in the aerated grit chambers and oxic tanks. Figure 5.10 shows the relationship between DO concentration in wastewater and methane flux in the aerated grit chambers and oxic tanks. The coefficients in the aerated grit chambers and oxic tanks were 0.772 and 0.656, respectively, so the correlation between methane flux and DO concentration was highly significant during the field experimental period. However, Aboobakar *et al.* (2014) reported that the aeration rate rather than DO concentration correlated well with CH_4 emissions.

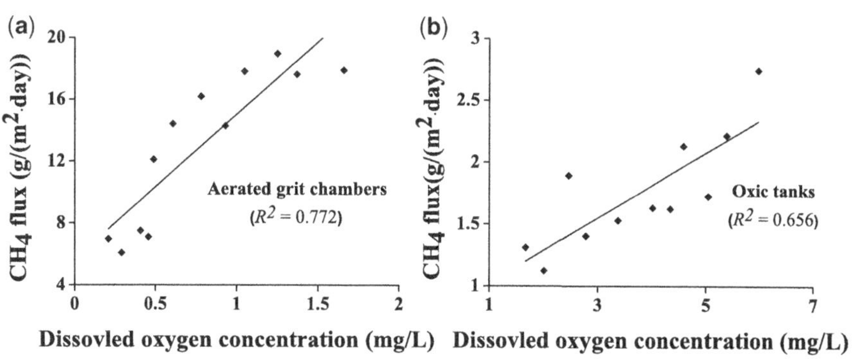

Figure 5.10 Relationship between CH_4 flux and dissolved oxygen concentration. (a) aerated grit chamber; (b) oxic tank (Wang *et al.* 2011a).

5.3.3 COD concentration and effect of C/N ratio

CH_4 generation in WWTPs is due to the biochemical conversion of COD to methane gas, so influent COD concentration in the wastewater and C/N ratio may affect CH_4 emission. Ren *et al.* (2015) investigated the influent C/N ratio on the CH_4 emission in a laboratory-scale A^2O bioreactor system. The A^2O bioreactor had a total working volume of 48 L with hydraulic retention time of 12 h and solid retention time of 12 days. The volume ratio of anaerobic tank/anoxic tank/oxic tank was 1:1:2. The mixed liquor suspended solids (MLSS) was maintained around 3500 ± 100 mg/L. The DO concentration in the oxic tanks was maintained around 2.5–3.0 mg/L. They set up two experiments to investigate the influence of C/N, as shown in Figure 5.11. In the first set, the C/N ratio was varied by changing the initial total nitrogen (TN) concentration, but keeping initial COD unchanged (Figure 5.11a). In the second set, the C/N ratio was varied by changing initial COD concentrations, but keeping initial TN unchanged (Figure 5.11b).

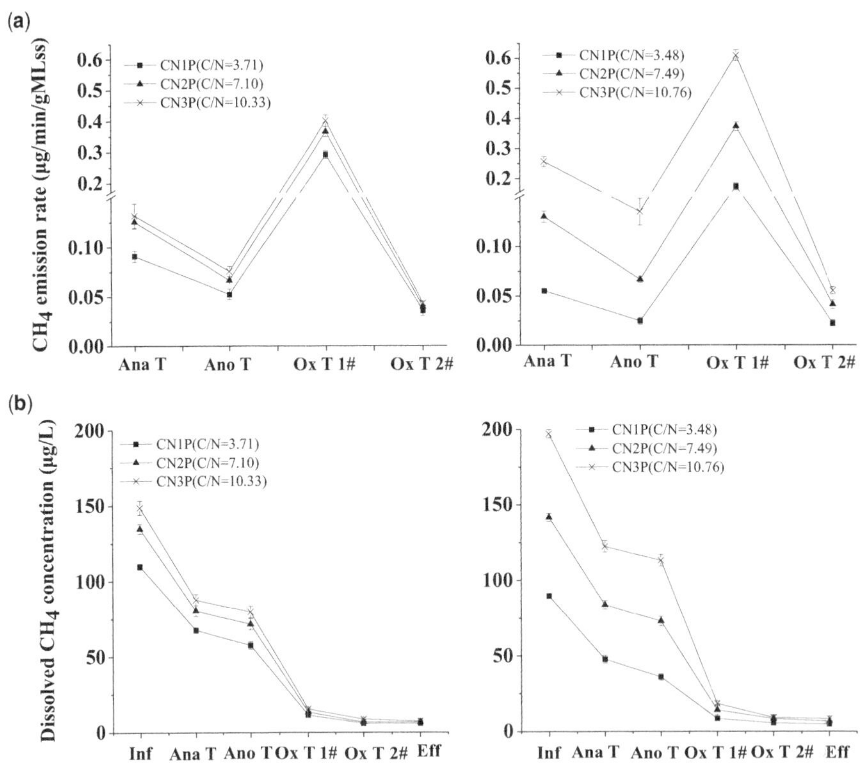

Figure 5.11 Effect of C/N ratio on the methane emission. (a) initial COD unchanged; (b) initial TN unchanged (Ren *et al.* 2015).

They found that the CH_4 emission rates in all operation units were mainly determined by the initial COD concentration in the influent, but not determined by the C/N ratio. The influence of C/N ratio on the dissolved CH_4 concentration had a similar trend to that of CH_4 emission. C/N ratio just slightly affected the dissolved CH_4 concentration, but the COD concentration markedly affected it.

5.4 METHANE OXIDATION

During the wastewater treatment process, a large proportion of CH_4 generated in the wastewater treatment process unit is oxidized in the activated sludge tanks. Daelman *et al.* (2012) reported that about 80% of the methane generated in the WWTP was biologically oxidized, while Liu *et al.* (2014) reported that 32.5–89.5% of CH_4 was biologically oxidized. Therefore, methane biological oxidation is an important pathway for mitigation of CH_4 emission from WWTPs. Methane oxidation can take place in both aerobic and anaerobic conditions (Malyan *et al.* 2016), which occurs via aerobic and anaerobic methanotrophs. Methane oxidation methanotrophs generally use CH_4 or methanol as a source of energy for their growth (Semrau *et al.* 2010). Aerobic methanotrophs (obligate methanotrophs) only utilize CH_4 as a carbon source and for energy, while anaerobic facultative methanotrophs grow on multi-carbon substrates (Dedysh & Dunfield, 2011).

5.4.1 Aerobic methane oxidation

In the aerobic CH_4 oxidation, CH_4 is oxidized into CO_2 by the actions of sequential enzymes. In the first step, methane monooxygenase (MMO) enzymes convert CH_4 into CH_3CHO, CH_3CHO is further oxidized to formaldehyde by methanol dehydrogenase, and subsequently formaldehyde is oxidized to formate and finally to CO_2 (Malyan *et al.* 2016) as described in Eq. 5.1:

$$CH_4 \rightarrow CH_3CHO \rightarrow HCHO \rightarrow HCOOH \rightarrow CO_2 \tag{5.1}$$

MMO enzymes catalyze the aerobic oxidation of CH_4. MMO are classified into two forms: particulate or membrane-bound form (pMMO) and soluble cytoplasmic form (sMMO) (Semrau *et al.* 2010), and both forms require oxygen for the oxidation of CH_4. The pMMO enzyme contains iron and copper, while sMMO is a cytoplasmic enzyme containing a unique di-iron site at its catalytic center (Chowdhury & Dick, 2013). The sMMO enzyme is capable of oxidizing a wide range of aliphatic, aromatic and alkane compounds (Malyan *et al.* 2016).

5.4.2 Anaerobic methane oxidation

The anaerobic oxidation of methane (AOM) is an important sink of methane in the atmosphere and significantly affects global warming. AOM is carried out by physical association of anaerobic methanotrophic (ANME) achaea and sulfate-reducing bacteria (SRB) (Nazaries *et al.* 2013). The SRB oxidizes CH_4 to CO_2 (reversed methanogenesis) by using sulfate as the electron acceptor, so the process is also called sulfate-dependent CH_4 oxidation. In 2006, a new AOM process named nitrite-dependent anaerobic methane oxidation (N-DAMO) was reported, in which nitrite is the electron acceptor of AOM (Raghoebarsing *et al.* 2006). Nitrate is also found to be an electron acceptor of AOM (Haroon *et al.* 2013). AOM is dependent on iron and manganese in the marine environment (Beal *et al.* 2009). Overall, there are three different processes of AOM depending on the different electron acceptors: sulfate-dependent anaerobic methane oxidation (S-DAMO), nitrate/nitrite-dependent anaerobic methane oxidation (N-DAMO), and metal ion (Mn^{4+} and Fe^{3+})-dependent anaerobic methane oxidation (M-DAMO), as shown in Figure 5.12 (Cui *et al.* 2015). All these processes may contribute to the oxidation of CH_4 in the anaerobic tank of a WWTP.

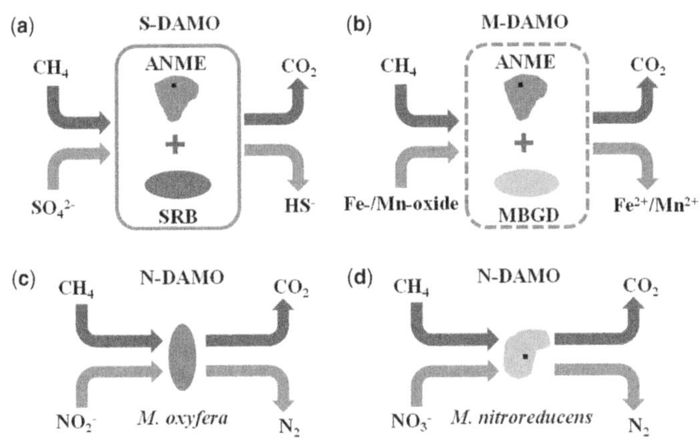

Figure 5.12 Three different models of anaerobic methane oxidation (AOM) depending on the different electron acceptors: (a) sulfate-dependent anaerobic methane oxidation (S-DAMO); (b) metal ion (Mn^{4+} and Fe^{3+})-dependent anaerobic methane oxidation (M-DAMO); and (c, d) nitrate/nitrite-dependent anaerobic methane oxidation (N-DAMO) (Cui *et al.* 2015).

5.4.3 Chemical oxidation of CH_4

Methane can also be oxidized at elevated temperature with the presence of catalysts, i.e., Pd/Co_3O_4-based catalysts, to form CO_2 and H_2O (Ercolino *et al.* 2015):

$$CH_4 + O_2 \rightarrow CO_2 + H_2O \tag{5.2}$$

Methane is able to be converted to CO and H_2 with the aid of catalysts, for example, Ni/MgO catalyst (Usman *et al.* 2015):

$$CH_4 + \frac{1}{2}O_2 \rightarrow CO + H_2 \tag{5.3}$$

There is also a possibility that CH_4 reacts with N_2O with the presence of catalysts to form methanol (Wood *et al.* 2004).

However, these chemical oxidation reactions of CH_4 need intensive inputs of energy and chemicals, which induces significant indirect GHG emissions. Therefore, it would not be practical to use chemical oxidation to mitigate direct CH_4 emission in WWTPs.

5.5 MITIGATION STRATEGIES FOR METHANE EMISSION IN WWTPs

Methane is a potent greenhouse gas, but it is also a clean energy source, and releases 8,590 kcal energy through complete combustion of 1 m^3 of methane (Demirbas, 2010). Methane emission mitigation in WWTPs can be achieved via three methods: methane collection, methane oxidation in aerobic tanks, and reduction of methane generation. Methane is mainly generated within sewer networks and anaerobic units, and is mainly emitted in the aeration units. Therefore, the off-gas in aeration units can be collected by covering these areas. The collected off-gas can be recirculated to the activated sludge systems where CH_4 can be oxidized. Based on the mass balance of methane in wastewater treatment processes, most of the generated methane is oxidized to carbon dioxide in aeration tanks (Wang *et al.* 2011a; Daelman *et al.* 2012a). Therefore, more investigations are needed on the impacts of the structure of WWTPs on methane emission, which can enhance methane oxidation and reduce the methane emission during wastewater treatment. More important, if WWTPs use the anaerobic digestion process to treat wastewater and waste sludge and recover methane gas for energy generation, this can significantly offset the greenhouse gas emission directly and indirectly in WWTPs.

5.5.1 Electricity generation from methane

In many countries, CH_4 generated from anaerobic digestion of wastewater (Li & Yu, 2011), sludge (Appels *et al.* 2008), and agricultural by-products (Xie *et al.* 2011a; Xie *et al.* 2011b; Xie *et al.* 2012), is used for electricity and heating production or as fuel. If anaerobic digestion is used as the wastewater treatment process for removing COD, the energy contained in wastewaters can be recovered in the form of CH_4. It has been predicted that the potential annual energy production from wastewater could reach 146.06 billion kWh by 2030. This reduces the use of grid electricity and natural gas, resulting in the indirect reduction in CO_2 emission. Due to the electricity generation from biogas, the indirect reduction in CO_2 emission in a wastewater treatment plant can be calculated using following equation (Oshita *et al.* 2014):

$$K_d = f_d \times p_d \times E/W \times 1000 \tag{5.4}$$

where

$K_d = CO_2$ equivalent per unit of influent wastewater (g CO_2 eq/m^3 influent);
f_d = annual biogas production in the wastewater treatment plant (m_N^3/year);
p_d = electricity generated from per unit volume of biogas using engine fueled with only biogas (1.88 kWh/m_N^3 is commonly used);
E = amount of CO_2 produced per unit of electricity generation using fossil fuels, which is country-specific (refer to Chapter 2); and
W = annual treatment capacity of the wastewater treatment plant (m^3 influent/year). For a certain wastewater treatment plant, the values of f_d and W are determined, and the indirection reduction in CO_2 emissions can be calculated.

For municipal WWTPs, anaerobic digesters are widely built to recover biogas from surplus sludge. According to the reports, the energy produced from anaerobic digestion of sewage sludge can provide approximately 60% of the energy consumed in WWTPs (Cao & Pawłowski, 2012). However, the contribution to the reduction of CO_2 emissions by the biogas generated varies from plant to plant. Oshita *et al.* (2014) obtained an indirect CO_2 emissions' reduction of 22.4% by electricity generation from biogas in two Japanese WWTPs, while according to Daelman *et al.* (2012), the value was 46.5% in a Netherlands WWTP. There was an obvious difference between the two studies, which may have been caused by the wastewater characteristics, wastewater treatment processes and anaerobic digester's efficiency. In the study of Oshita *et al.* (2014), the specific methane yield (SMY) of sludge in the digester was only about 50% of the normal level.

However, methane gas leaks from the anaerobic digesters and slips away in gas engines. The amount of methane slip of the gas engine can be up to 1.3–1.8% (Daelam *et al.* 2012). Due to its high global warming potential (GWP) values, the methane slip would lessen the reduction of carbon footprints achieved by using the biogas. Daelman *et al.* (2012) showed that in Kralingseveer WWTP, the methane emitted from the anaerobic digester was 5.7 tonnes CO_2 equivalent, exceeding the reduction of carbon footprint due to burning of the biogas. Hence, proper design and operation of the anaerobic digesters and an efficient use of methane is necessary; otherwise, the anaerobic digesters would be potent GHG producers.

5.6 SUMMARY

Reducing methane emission in wastewater treatment processes can directly mitigate the greenhouse gas emission. Anaerobic conditions in wastewater treatment result in the production of CH_4 and potentially increase the CH_4 emission. Aerobic conditions inhibit CH_4 generation and enhance the oxidation of CH_4, but increase the emission. However, aeration requires air supply by air blowers or mechanical aerators, which means more electricity consumption. If electricity generation for the operation of WWTPs is from consuming fossil fuels, this also leads to increased indirect GHGs emission.

Methane can be utilized to generate electricity, heat wastewater or sludge digestion tanks, and drive vehicles. Therefore, collecting and recovering CH_4 produced during wastewater treatment and anaerobic digestion of sewage sludge can significantly mitigate the greenhouse gas emission from wastewater treatment processes.

5.7 REFERENCES

Aboobakar A., Jones M., Vale P., Cartmell E. and Dotro G. (2014). Methane emissions from aerated zones in a full-scale nitrifying activated sludge treatment plant. *Water, Air & Soil Pollution*, **225**(1), 1814.

Appels L., Baeyens J., Degreve J. and Dewil R. (2008). Principles and potential of the anaerobic digestion of waste-activated sludge. *Progress in Energy and Combustion Science*, **34**(6), 755–781.

Beal E. J., House C. H. and Orphan V. J. (2009). Manganese- and iron-dependent marine methane oxidation. *Acta Agriculturae Scandinavica Section A Animal Science*, **325**(5937), 184–187.

Cao Y. and Pawłowski A. (2012). Sewage sludge-to-energy approaches based on anaerobic digestion and pyrolysis: brief overview and energy efficiency assessment. *Renewable and Sustainable Energy Reviews*, **16**(3), 1657–1665.

Chowdhury T. R. and Dick R. P. (2013). Ecology of aerobic methanotrophs in controlling methane fluxes from wetlands. *Applied Soil Ecology*, **65**, 8–22.

Cui M. M., Ma A. Z., Qi H. Y., Zhuang X. L. and Zhuang G. Q. (2015). Anaerobic oxidation of methane: an "active" microbial process. *MicrobiologyOpen*, **4**(1), 1–11.

Czepiel P. M., Crill P. M. and Harriss R. C. (1993). Methane emissions from municipal wastewater treatment processes. *Environmental Science & Technology*, **27**(12), 2472–2477.

Daelman M. R., van Voorthuizen E. M., van Dongen U. G., Volcke E. I. and van Loosdrecht M. C. (2012). Methane emission during municipal wastewater treatment. *Water Research*, **46**(11), 3657–3670.

Dedysh S. N. and Dunfield P. F. (2011). 3 facultative and obligate methanotrophs: how to identify and differentiate them. *Method Enzymol*, **495**, 31.

Demirbas A. (2010). Methane hydrates as potential energy resource: Part 1-Importance, resource and recovery facilities. *Energy Conversion and Management*, **51**(7), 1547–1561.

Ercolino G., Grzybek G., Stelmachowski P., Specchia S., Kotarba A. and Specchia V. (2015). Pd/Co3O4-based catalysts prepared by solution combustion synthesis for residual methane oxidation in lean conditions. *Catalysis Today*, **257**, 66–71.

Haroon M. F., Hu S., Shi Y., Imelfort M., Keller J., Hugenholtz P., Yuan Z. and Tyson G. W. (2013). Anaerobic oxidation of methane coupled to nitrate reduction in a novel archaeal lineage. *Bionature*, **500**(7464), 567.

Jin L. Y., Zhang G. M. and Tian H. F. (2014). Current state of sewage treatment in China. *Australian Journal of Marine and Freshwater Research*, **66**, 85–98.

Li W. W. and Yu H. Q. (2011). From wastewater to bioenergy and biochemicals via two-stage bioconversion processes: a future paradigm. *Biotechnology Advances*, **29**(6), 972–982.

Liu Y., Cheng X., Lun X. and Sun D. (2014). CH_4 emission and conversion from A^2O and SBR processes in full-scale wastewater treatment plants. *Journal of Environmental Sciences*, **26**(1), 224–230.

Liu Y. W., Ni B. J., Sharma K. R. and Yuan Z. G. (2015). Methane emission from sewers. *Science of the Total Environment*, **524**, 40–51.

Liu Y. W., Tugtas A. E., Sharma K. R., Ni B. J. and Yuan Z. G. (2016). Sulfide and methane production in sewer sediments: field survey and model evaluation. *Australian Journal of Marine and Freshwater Research*, **89**, 142–150.

Malyan S. K., Bhatia A., Kumar A., Gupta D. K., Singh R., Kumar S. S., Tomer R., Kumar O. and Jain N. (2016). Methane production, oxidation and mitigation: a mechanistic understanding and comprehensive evaluation of influencing factors. *Science of the Total Environment*, **572**, 874–896.

Masuda S., Suzuki S., Sano I., Li Y. Y. and Nishimura O. (2015). The seasonal variation of emission of greenhouse gases from a full-scale sewage treatment plant. *Chemosphere*, **140**, 167–173.

Nazaries L., Murrell J. C., Millard P., Baggs L. and Singh B. K. (2013). Methane, microbes and models: fundamental understanding of the soil methane cycle for future predictions. *Applied and Environmental Microbiology*, **15**(9), 2395–2417.

Oshita K., Okumura T., Takaoka M., Fujimori T., Appels L. and Dewil R. (2014). Methane and nitrous oxide emissions following anaerobic digestion of sludge in Japanese sewage treatment facilities. *Bioresource Technology*, **171**, 175–181.

Raghoebarsing A. A., Pol A., van de Pas-Schoonen K. T., Smolders A. J. P., Ettwig K. F., Rijpstra W. I. C., Schouten S., Damste J. S. S., Op den Camp H. J. M., Jetten M. S. M. and Strous M. (2006). A microbial consortium couples anaerobic methane oxidation to denitrification. *Bionature*, **440**(7086), 918–921.

Ren Y., Wang J., Xu L., Liu C., Zong R., Yu J. and Liang S. (2015). Direct emissions of N_2O, CO_2, and CH_4 from A/A/O bioreactor systems: impact of influent C/N ratio. *Environmental Science and Pollution Research*, **22**(11), 8163–8173.

Semrau J. D., DiSpirito A. A. and Yoon S. (2010). Methanotrophs and copper. *FEMS Microbiology Reviews*, **34**(4), 496–531.

Usman M., Daud W. and Abbas H. F. (2015). Dry reforming of methane: influence of process parameters – a review. *Renewable & Sustainable Energy Reviews*, **45**, 710–744.

Wang J., Zhang J., Xie H., Qi P., Ren Y. and Hu Z. (2011). Methane emissions from a full-scale A/A/O wastewater treatment plant. *Bioresource Technology*, **102**(9), 5479–5485.

Wood B. R., Reimer J. A., Bell A. T., Janicke M. T. and Ott K. C. (2004). Methanol formation on Fe/Al-MFI via the oxidation of methane by nitrous oxide. *Journal of Catalysis*, **225**(2), 300–306.

Xie S., Frost J. P., Lawlor P. G., Wu G. and Zhan X. (2011a). Effects of thermo-chemical pre-treatment of grass silage on methane production by anaerobic digestion. *Bioresource Technology*, **102**(19), 8748–8755.

Xie S., Lawlor P. G., Frost J. P., Hu Z. and Zhan X. (2011b). Effect of pig manure to grass silage ratio on methane production in batch anaerobic co-digestion of concentrated pig manure and grass silage. *Bioresource Technology*, **102**(10), 5728–5733.

Xie S., Wu G., Lawlor P. G., Frost J. P. and Zhan X. (2012). Methane production from anaerobic co-digestion of the separated solid fraction of pig manure with dried grass silage. *Bioresource Technology*, **104**, 289–297.

Zhang Q. H., Yang W. N., Ngo H. H., Guo W. S., Jin P. K., Dzakpasu M., Yang S. J., Wang Q., Wang X. C. and Ao D. (2016). Current status of urban wastewater treatment plants in China. *Environment International*, **92–93**, 11–22.

Chapter 6

N_2O emission during biological nitrogen removal

Guangxue Wu, Bo Li and Yan Yang

6.1 OVERVIEW OF NITROGEN CYCLE

The schematic diagram of the nitrogen cycle during wastewater treatment is shown in Figure 6.1. It includes oxidization and reduction processes, with the redox state of N varying from –3 (in ammonium, NH_4^+) to +5 (in nitrate, NO_3^-). Conventional biological nitrogen removal (BNR) from municipal and domestic wastewater is mainly through autotrophic nitrification and heterotrophic denitrification processes. During nitrification, NH_4^+ is sequentially oxidized to nitrite (NO_2^-) and NO_3^-, with production of by-products of nitrogen oxide (NO) and nitrous oxide (N_2O). During denitrification, which is contrary to nitrification, oxidized nitrogen (NO_2^- and NO_3^-) is reduced to nitrogen (N_2) gas. Other nitrogen cycle processes include anaerobic ammonium oxidation (Anammox), ammonification and N fixation.

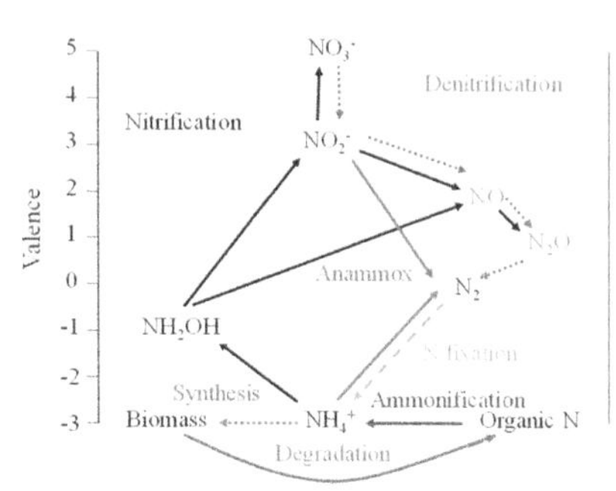

Figure 6.1 Processes within the nitrogen cycle which occur during wastewater treatment.

Xinmin Zhan, Zhenhu Hu, Guangxue Wu
doi: 10.2166/9781780406312_101

6.2 IDENTIFICATION OF N_2O EMISSION PATHWAYS IN BNR

N_2O is emitted from both nitrification and denitrification through different pathways (Section 1.5, Chapter 1). To date, the methods to differentiate N_2O emission pathways are still under development.

Since all pathways of N_2O emissions are related to microbial activities, chemical inhibitors which inhibit certain nitrification or denitrification biochemical reactions are being tried to determine N_2O generation pathways. The commonly applied chemicals include allylthiourea for inhibiting NH_4^+ oxidizing bacteria (AOB), chlorate for inhibiting NO_2^- oxidizing bacteria (NOB), and chlorate for inhibiting denitrification from NO_2^- to N_2 (Chen *et al.* 2014). However, the accuracy of these methods needs to be assessed. The question of the reliability of inhibition by acetylene and oxygen on nitrifier denitrification and nitrification has been raised by Wrage *et al.* (2004). When acetylene (10 kPa) is used as an inhibitor to inhibit N_2O reduction, under carbon limited conditions, this chemical is also used by denitrifiers as the carbon substrate for denitrification (Groffman *et al.* 2006).

In addition to chemical inhibitors, some new technologies have also been investigated to differentiate N_2O emission pathways: microsensor measurements, isotopic composition analysis and also microbial community analysis (including functional gene detected by real time polymerase chain reaction, RT-PCR) (Schreiber *et al.* 2012). The isotopic measurement technique can identify if N_2O emission is by AOB (Peng *et al.* 2014) or due to denitrification. Mathematical modeling helps to clarify mechanisms responsible for the N_2O emission, and differentiate oxidation pathways and nitrifier denitrification pathways (Ni *et al.* 2014). A metabolic network model can clarify the regulation mechanisms of NO and N_2O production by AOB (Perez-Garcia *et al.* 2014).

It is expected that N_2O emission mainly occurs in the aerobic phase of intended nitrification and denitrification systems due to the air stripping effect. For instance, Lim and Kim (2014) found that the aerobic N_2O emission accounted for 95% of the total emission.

A survey from 12 full-scale wastewater treatment plants (WWTPs) confirms that the aerated tanks are the main source of N_2O emission (Ahn *et al.* 2010). In a sequencing batch reactor (SBR), the N_2O emission factor was 8.6–16.1% during the aerobic phase and only 0–0.05% during the anoxic phase (Jian *et al.* 2006).

6.3 FACTORS AFFECTING N_2O EMISSION DURING NITRIFICATION

During nitrification, N_2O emission is affected by substrates such as NH_4^+, O_2 and NO_2^-, and also environmental factors affecting activities of nitrifiers such as metals and pH etc., as shown in Figure 6.2. Understanding how these factors impact N_2O emission will help us to mitigate its emission in WWTPs.

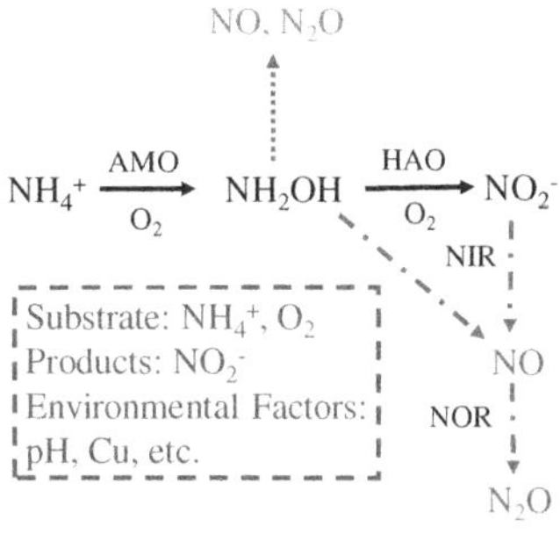

Figure 6.2 Factors affecting N_2O emission during nitrification. AMO: Ammonia monooxygenase; HAO: hydroxylamine dehydrogenase; NIR: nitrite reductase; NOR: nitric oxide reductase.

6.3.1 Dissolved oxygen

Nitrification requires O_2 as the electron acceptor, so nitrifying tanks are aerated. WWTP operators like using a low dissolved oxygen (DO) concentration in the nitrifying tanks so as to reduce energy input, while this practice could lead to a high N_2O emission.

At low DO conditions, nitrifiers can use NO_2^- rather than O_2 as the electron acceptor; this induces high N_2O emissions. DO concentrations below 2 mg/L can cause significant N_2O emissions in nitrifying tanks and N_2O emission would be negligible when DO is higher than 2 mg/L. Hence, N_2O emission factors in the nitrifying tanks depend on the DO concentrations. It is recommended to maintain a 2 mg/L DO in the nitrifying tank, at least 1 mg/L, so as to reduce its emission. According to the activated sludge floc structure (Figure 6.3), when the bulk fluid has a low DO level, the DO concentrations inside the flocs would be low enough for the denitrifiers, which exist in the center of the flocs, to reduce oxidized nitrogen to N_2, leading to production of some N_2O as by-products and consequently a high N_2O emission factor. Tallec *et al.* (2006) found that in nitrifying conditions, the highest N_2O emission was observed at a DO of 1 mg/L; when DO was in the range of 0.1 to 2 mg/L, nitrifier denitrification contributed 58–83% of the total N_2O emission; the N_2O emission factor to the oxidized NH_4-N was in the range of 0.1–0.4% depending on the DO concentration.

DO concentrations impact N_2O emissions together with other operating conditions. One of these is the nitrogen loading rates (Aboobakar *et al.* 2013). A high N_2O emission is more likely to happen under high NH_4^+ and influent nitrogen loading rates. A high NH_4^+ loading rate induces a low DO concentration in the nitrifying tank and then a high N_2O emission. A high N_2O emission may be monitored when NH_4^+ is used up in nitrifying tanks, which could be due to the fact that some biochemical reactions using NO_3^- induce the N_2O emission.

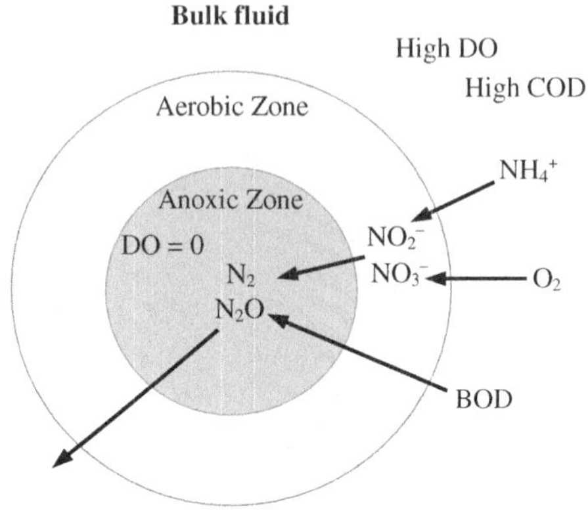

Figure 6.3 Activated sludge floc structure. COD: chemical oxygen demand; BOD: biochemical oxygen demand

Another factor is sludge retention time (SRT). A short SRT and a low DO concentration (Zheng *et al.* 1994) would promote a high N_2O emission. A short SRT causes incomplete nitrification and high accumulation of NO_2^-, which would increase N_2O emission. The highest N_2O emission factor would be up to 16% at a SRT of 3 days. Decreasing SRT increases N_2O emissions for both nitrification and denitrification processes (Noda *et al.* 2004).

Changing operating conditions from anoxic to aerobic conditions, or from aerobic to anoxic conditions will cause a shift of the nitrifiers' (and denitrifiers') metabolism from a low specific activity towards the maximum specific activity or from the maximum specific activity to a lower one; during this shift, N_2O emission is increased (Yu *et al.* 2010). For instance, Kampschreur *et al.* (2008) found that N_2O emission under steady state was 2.8%, while this was increased under dynamic conditions with transit dynamics of NH_4^+, NO_2^- and O_2. Rassamee *et al.* (2011) examined N_2O emission under fully aerobic nitrifying conditions and intermittently aerobic conditions, and found that aerobic N_2O emission only occurred when both low DO (below 4 mg/L) and high NO_2^- (above 1.5 mg/L) coexisted; changes in DO, NO_2^- and NH_4^+ promoted N_2O emission. The N_2O emission factor to the total nitrogen (TN) was 0.17–0.24% for fully aerobic conditions, and 0–0.39% for intermittently aerobic conditions.

Activated sludge floc structure also impacts the effect of DO on nitrification and N_2O emission because it affects the distribution of DO in the flocs. Hence, N_2O emission might be also dependent on the cell density and different initial cell concentrations (Remde & Conrad, 1990; Wrage *et al.* 2004).

6.3.2 NO_2^-

During nitrification, the N_2O emission is dependent on the oxidation of NH_4^+ by AOB (Hu *et al.* 2010; Kim *et al.* 2010). Under O_2 limited conditions, NO_2^- can be denitrified by AOB to N_2O or N_2 with ammonia, hydrogen or other organic carbons as the electron donor. Poth *et al.* (1985) for the first time confirmed that nitrifying bacteria can use NO_2^- as the electron acceptor to generate N_2O; this process is now known as nitrifier denitrification. This can be confirmed with an experiment designed for nitrifying systems: if NH_4^+ is completely eliminated from the wastewater treatment system, the emission of N_2O will be stopped but will continue with the addition of NO_2^- (Kim *et al.* 2010).

The existence of NO_2^- in the nitrifying systems would increase the N_2O emission. In their nitrifying laboratory-scale experiments, where nitrifier denitrification represented 58–83% of the total N_2O production, Tallec *et al.* (2006) observed that adding NO_2^- into the systems always significantly increased N_2O emission and the increase was dependent on the DO levels: four times higher at 0.1 and 2 mgO_2/L, six times at 0.6 mgO_2/L and eight times at 1 mg O_2/L.

Just as Figure 6.3 indicates, even in the nitrifying tanks, heterotrophic denitrification can also occur. The addition of NO_2^- increases N_2O emission not only through nitrifier denitrification but also through heterotrophic denitrification. How NO_2^- impacts the heterotrophic denitrification pathway will be described in Section 6.4.2.

6.3.3 pH

Like other microorganisms, the activities of nitrifiers and denitrifiers are affected by pH, which can directly or indirectly affect N_2O emission. In addition, pH determines the concentrations of free nitrous acid (FNA) and free ammonia (FA) (Eqs. 6.1, 6.2, 6.3 and 6.4); both parameters affect the N_2O emissions directly (Zhou *et al.* 2008).

$$NH_4^+ + OH^- \leftrightarrow NH_3 + H_2O \tag{6.1}$$

$$FA = \frac{[TAN] \times 10^{pH}}{K_b/K_w + 10^{pH}} \times \frac{17}{14} \tag{6.2}$$

where

$[TAN]$ = NH_4^+-N concentration;

K_b = ionization constant of the ammonia equilibrium; and

K_w = ionization constant of water. $K_b/K_w = \exp(6334/(273 + T))$, and T = temperature, °C.

$$H^+ + NO_2^- \leftrightarrow HNO_2 \tag{6.3}$$

$$FA = \frac{[NO_2^- - N] \times 10^{-pH}}{K_a + 10^{-pH}} \times \frac{47}{14} \tag{6.4}$$

where

$[NO_2^-\text{-}N]$ = NO_2^--N concentration; and

K_a = ionization constant of the nitrous acid equilibrium. $K_a = \exp(-2300/(273 + T))$, and T = temperature, °C.

Law *et al.* (2011) examined the effects of pH on N_2O production for a partial nitrifying system, and found that the N_2O emission factor was around 1% of the NH_4^+ converted. The N_2O emission was low at pH of 6–7 and high at pH 8. For municipal WWTPs, pH is quite stable between 6 and 8, so the effect of pH on N_2O emission is limited.

6.4 OPERATING FACTORS AFFECTING N_2O EMISSION DURING DENITRIFICATION

Denitrification is sequential reduction of oxidized nitrogen to nitrogen gas, with different enzymes associated. The factors affecting N_2O emission during denitrification include electron donors (organic carbon), electron acceptors (NO_2^- and NO_3^-) and the applied environmental conditions, such as pH and SRT etc. Many parameters such as substrate concentrations, pH, NO_2^- accumulation and so on altogether influence the N_2O emission during denitrification (Figure 6.4).

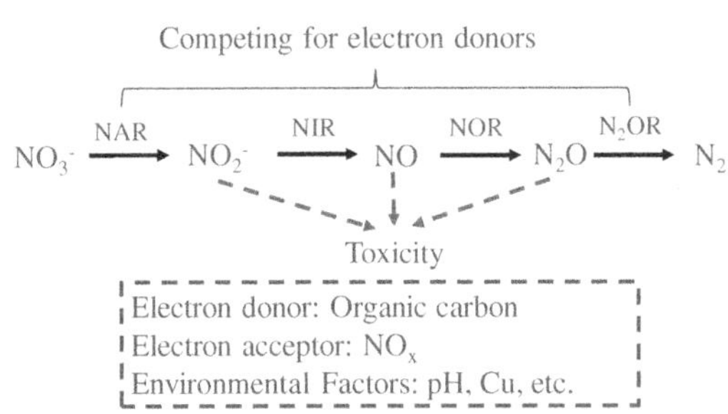

Figure 6.4 Factors affecting N_2O emission during denitrification.

6.4.1 Organic carbon (types and C/N ratios)

Among all parameters, both the types of organic matter in the wastewater and the C/N ratios have a significant effect on the N_2O emission during heterotrophic denitrification. When incomplete denitrification occurs, a high N_2O emission potential would exist. COD/N is an important parameter that can influence the denitrification efficiency; a high C/N ratio enables complete denitrification, and the end product of denitrification is N_2. Under a low C/N ratio, limited availability of biodegradable organic carbon tends to increase N_2O emission during denitrification. During denitrification in WWTPs, low COD/N ratios causing N_2O accumulation are reported to be between 3.5 and 1.5 (Hanaki *et al.* 1992; Itokawa *et al.* 1996; Chung & Chung, 2000; Kishida *et al.* 2004). When external carbon is insufficient, denitrifiers will denitrify with internal carbon sources or endogenous respiration, for example, polyhydroxyalkanoates (PHA) and polyhydroxybutyrate (PHB). In this case, there is a clear positive correlation between PHA/PHB utilization and N_2O emissions (Figure 6.5; Pan *et al.* 2012).

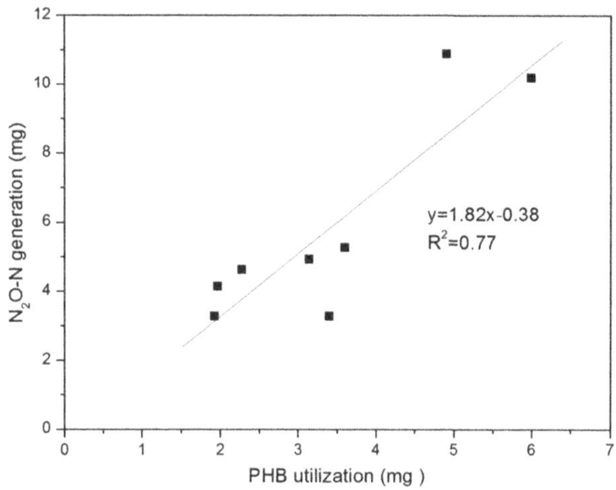

Figure 6.5 Relationship between PHB utilization and N_2O generation in an intermittently aerated sequencing batch reactor treating slaughterhouse wastewater at 11°C (Pan *et al.* 2012).

On the other hand, limited carbon conditions deteriorate denitrification and cause the accumulation of NO_2^-, which further inhibits N_2O reduction, resulting in N_2O emission. Alinsafi *et al.* (2008) obtained the N_2O emission factor of 5.1%, 2.6% and 1% at COD/N ratios of 3, 5 and 7. Kishida *et al.* (2004) found that the N_2O emission factor was 1.71% at BOD/N of 4.5 and 17.7% at 2.6.

However, this opinion is not shared by some researchers due to their findings. Low COD/N ratios could potentially intensify the electron competition among the denitrification reductases through slowing down the electron supply rates. The intensity of electron competition is likely to govern N_2O accumulation during denitrification and electron competition can occur under both carbon limiting and abundant conditions (Pan *et al.* 2013). Therefore, it is not necessarily true that a low COD/N ratio alone leads to N_2O accumulation. It has been observed that there is no N_2O accumulation under carbon limited conditions during NO_3^- reduction by enriched methanol or ethanol utilizing denitrifiers, even when the COD/N ratio was as low as 1.5 (Lu & Chandran, 2010; Pan *et al.* 2013). This may be because of the pure cultures assessed in these studies, which are not representative of real WWTPs.

The effect of organic carbon limitation on N_2O emission also depends on the acclimation history and the operating conditions (Hanaki *et al.* 1992; Lu & Chandran, 2010). The types of organic carbon have a high influence on denitrifier communities acclimated and their distinct metabolic pathways. When adding easily biodegradable organic matter to the denitrifying systems, like acetate, methanol, glucose, etc., N_2O emission from the WWTPs would be reduced.

Belmonte *et al.* (2012) observed that there was no N_2O emission with acetate as the organic carbon, while there was a high emission with swine wastewater as the organic carbon, and its emission factor increased with increasing the denitrifying rate.

Impacts of COD/N ratios on denitrifying systems are also enhanced with other WWTP operation factors, such as SRT. A low COD/N ratio and short SRT promote a high N_2O emission due to the incomplete denitrification.

6.4.2 NO_2^--N

N_2O is the intermediate during denitrification, so incomplete denitrification leads to N_2O emission. The possible reasons for the emission of N_2O during denitrification are as follows: (1) the accumulation of NO_2^- limits the activity of N_2O reductase (NOS) (Alinsafi *et al.* 2008), leading to the emission of N_2O; (2) some denitrifiers do not have NOS, such as *Paracoccus denitrificans* (Takaya *et al.* 2003). The final product of these denitrifiers is N_2O instead of N_2; and (3) the competitiveness of NOS for electron donors is weak compared with other enzymes associated with denitrification (Schalk-Otte *et al.* 2000). During denitrification, high concentrations of NO_2^- promote the release of N_2O. Especially, nowadays, partial nitrification is widely studied and applied with NO_2^- as an important intermediate substance. Even though this process is able to save energy, it may increase N_2O emission. NO_2^- and NO can inhibit NO_3^- or N_2O reductases because they have high affinity for metal ions at the active sites of enzymes. In addition, the toxicity of NO and FNA may also inhibit N_2O reduction (Alinsafi *et al.* 2008).

Based on laboratory-scale denitrifying systems, Alinsafi *et al.* (2008) observed that the emission of N_2O increased with pulse addition of NO_2^- and the highest nitrogen conversion to N_2O gas (14.4%) was recorded in an influent wastewater containing both NO_2^- and NO_3^-. In a simultaneous nitrification and denitrification (SND) and phosphorus removal process, when NO_2^- was used as the electron acceptor, the N_2O emission factor was 77%, while it was 26% with NO_3^- as the electron acceptor (Lemaire *et al.* 2006).

Pulse addition of NO_2^- may not necessarily induce a high N_2O emission, which is also dependent on the acclimation history and the applied organic carbons (Lu & Chandran, 2010). Hanaki *et al.* (1992) found that the inhibition concentration of NO_2^- was 10 mg/L for acetate and yeast acclimated denitrifiers; while with sucrose as the sole organic carbon for denitrifiers, N_2O emission was not observed at the NO_2^- concentration of 10 mg/L (Thörn & Sörensson, 1996).

6.4.3 DO

N_2O emission is more affected by DO than NO_2^- due to the high inhibition of the NOS enzyme by DO (Von Schulthess *et al.* 1994). N_2O reductase is the most sensitive enzyme to oxygen (Bonin *et al.* 1989). Hence, the N_2O reduction enzyme may be inactivated when exposed to a sudden change in DO, while other enzymes are still active, inducing a high N_2O emission. Therefore, intermittent exposure to O_2 may cause a high N_2O emission. DO affects denitrification by two means: enzyme synthesis repression and enzyme activity inhibition. The former occurs within hours and its decay is slow, while the latter may be within minutes depending on the inhibition degree (Von Schulthess *et al.* 1994). By mathematical modeling, von Schulthess *et al.* (1994) stated that N_2O emission during denitrification was mainly due to the slower rate of N_2O reduction than those of NO_3^- and NO_2^- reduction. A DO concentration above 3.8 mg/L has been shown to inhibit denitrification (Beline *et al.* 2001).

The effect of DO on denitrification also depends on the acclimated denitrifiers, probably due to their different denitrification pathways. Lu and Chandran (2010) found that for ethanol acclimated denitrifiers exposed to oxygen, a high N_2O emission factor of 7.1% was recorded, while the value was low for methanol acclimated denitrifiers; the high DO inhibited NO_3^- reduction in the methanol-acclimated denitrifiers and resulted in a low denitrification rate, while the N_2O reduction of ethanol-acclimated denitrifiers was inhibited with no NO_3^- reduction.

At low DO concentrations (0.4–1.1 mg/L), denitrification may occur simultaneously with nitrification, so N_2O emission by heterotrophic denitrification may also take place at the same time by nitrifiers. If the nitrifying tank has a low DO concentration, high NO_2^- concentrations produced during nitrification would be reduced by heterotrophic denitrification; this would help to control high N_2O emission (Kim & Kim, 2011).

6.4.4 pH

pH affects N_2O emission during denitrification directly or indirectly. pH in the range of 7–9 does not increase N_2O emission but when pH drops to 6.5, higher N_2O emission will be induced. Low pH inhibits N_2O reductase, which is responsible for the reduction of N_2O to N_2 (Glass & Orphan, 2012). N_2O reductase is more active at pH above 7, while NO_3-N, NO_2-N and NO reductases are more active at pH below 7 (Richardson *et al.* 2009); this would lead to competition for electrons among these nitrogen oxide reductases at low pH conditions (Pan *et al.* 2012), causing N_2O accumulation and emission. In addition, when a high concentration of NO_2^- exists in the wastewater treatment system, lowering pH leads to high concentrations of FNA. HNO_2 inhibits N_2O reduction, which can take place at HNO_2 as low as 0.004 mg/L (Zhou *et al.* 2008). As a consequence, lowering pH would lead to N_2O emissions.

Therefore, close to neutral or higher pH, 6.8–8.5, does not lead to high N_2O emission in heterotrophic denitrification while when pH is acidic, below 6.5, N_2O emission is much higher (Hanaki *et al.* 1992).

6.4.5 Cu

Under oxygen limited conditions, reduction enzymes for nitrogen oxides are metalloenzymes located in the cell membrane and periplasm. The reduction of NO_2 to NO is catalyzed either by Fe-containing nitrite reductases (e.g., *Paracoccus denitrificans*) or by Cu-containing nitrite reductases (e.g., *Achromobacter xylosoxidans*), while the reduction of N_2O to N_2 is catalyzed only by Cu-containing N_2O reductases of denitrifiers (Richardson *et al.* 2009). As a consequence, the concentration of Cu in wastewater affects the formation and emission of N_2O. In some cases, supply of adequate Cu is necessary and important for the reduction of N_2O through activating the N_2O reductases. Too low Cu increases N_2O formation and emission. Increasing the NO_3^- loading rate enhances N_2O emission under Cu-free conditions and the emission would be reduced by adding a small amount of Cu (Ito & Matsumura, 2001). Zhu *et al.* (2013) observed that a Cu concentration of 10–100 μg/L reduced N_2O production by 54.7–73.2% through improving NO_2^- and N_2O reduction activities. Similar to adding Cu salts, adding Cu nanoparticles can improve the denitrification efficiency, reduce NO_2^- accumulation, and reduce N_2O emissions (Chen *et al.* 2012). For example, the N_2O generation was 0.441, 0.404, 0.360, 0.174 and 0.269 mg/mg N removed at the dosage of Cu nanoparticles of 0, 0.1, 1, 5 and 10 mg/L in wastewater (Chen *et al.* 2012).

While, effects of Cu on N_2O emission are dependent on types of denitrifiers. For example, *Paracoccus denitrificans*, whose NO_2^- reductases are Fe-containing, under NO_3^- sufficient conditions, released 40% of the reduced NO_3^- as N_2O under Cu deficient conditions. *Achromobacter xylosoxidans*, whose nitrite reductases are Cu-containing, released 40% of NO_3^- reduced as NO_2^-. Under Cu efficient and nitrate limited conditions, for both phenotypes, little N_2O or nitrite was detected (Felgate *et al.* 2012).

On the other hand, if Cu concentrations are too high, the activity of N_2O reduction will be inhibited, which induces high emissions of N_2O (Granger & Ward, 2003).

6.5 N_2O EMISSION FROM DIFFERENT BIOLOGICAL WASTEWATER TREATMENT PROCESSES

6.5.1 Conventional biological nitrogen removal processes

In conventional BNR processes, nitrogen removal is achieved through sequential nitrification and denitrification. It is quite common that denitrification is arranged prior to nitrification (Figure 6.6).

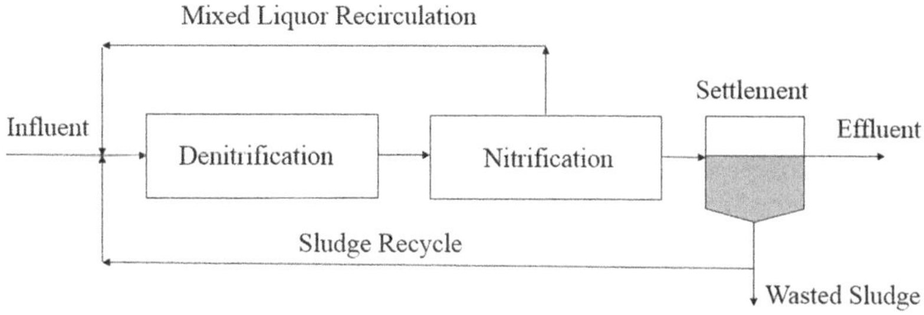

Figure 6.6 Conventional denitrification and nitrification processes for nitrogen removal.

Other widely used conventional BNR processes are the anaerobic/anoxic/oxic (A^2O) process (Figure 2.4, Chapter 2) and sequencing batch reactor (SBR) process, whose operation is time-based and in cyclic sequences, each consisting of fill, react, settle, draw and idle phases (Figure 6.7) repeated over time.

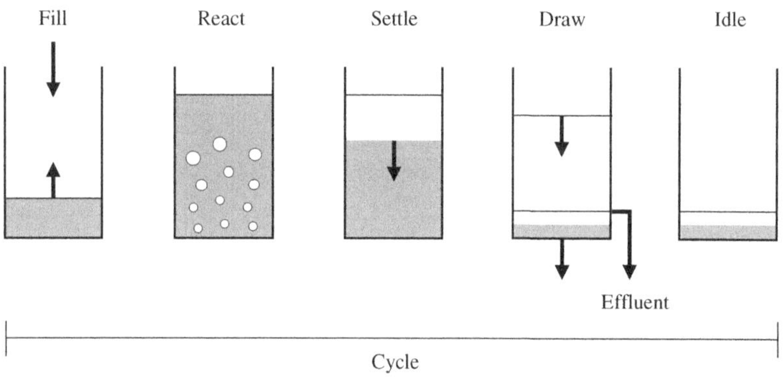

Figure 6.7 Time sequence of a conventional sequencing batch reactor.

In conventional BNR processes, N_2O emission mainly occurs during the aerobic phase due to gas stripping. Increasing NH_4^+, NO_2^- and temperature would increase N_2O emission (Gejlsbjerg *et al.* 1998). For N_2O emission from A^2O and SBR processes, Sun *et al.* (2013) found that the N_2O emission factor was 6.52% in SBR and 1.95% in A^2O, and both were mainly from the aerobic phase, with the contribution ratio of above 96.9% in A^2O and 99.9% in SBR. In the anoxic/oxic (A/O), A^2O and conventional SBR processes, the dissolved N_2O produced during the aerobic phase could be reduced to N_2 during the anoxic phase.

Anoxic to aerobic phase ratio (PR) is critical to N_2O emission. A low PR leads to incomplete denitrification and NO_2^- accumulation, inducing a high N_2O emission during nitrification; at a high PR, organic carbon is stored as cell internal organic compounds during the anoxic phase and consequently used for simultaneous nitrification and denitrification during the aerobic phase, also inducing a high N_2O emission factor. Hu *et al.* (2011a) obtained a low N_2O emission at PR of 0.5, with a N_2O emission factor of 9.8%.

The N_2O emission factors are in the range of 0.006–0.253 g/g denitrified N, with the average value of 0.035 g/g denitrified N (Foley *et al.* 2011).

6.5.2 Multiple A/O processes

The intermittently aerated multiple A/O process is widely applied nowadays for enhancing biological nitrogen removal. This process can be arranged as several sequential anoxic and aerobic phases (Figure 6.8) or be arranged by splitting an SBR's react phase into alternating aeration and non-aeration phases, as shown in Figure 6.9. By this means, nitrogen removal can be enhanced through staged anoxic and aerobic phases by better utilization of organic carbon in wastewater.

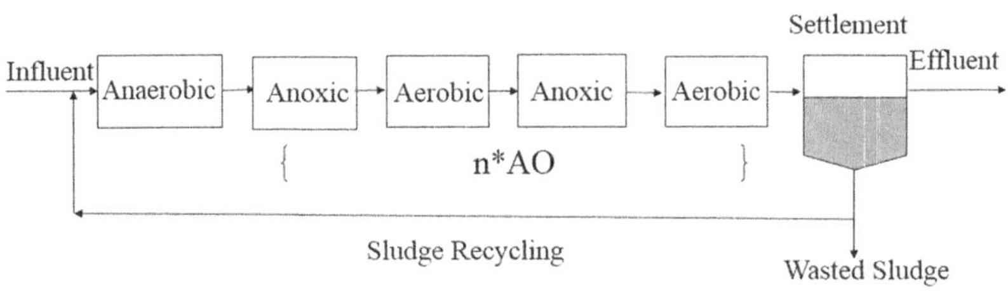

Figure 6.8 Multiple denitrification and nitrification processes for enhanced biological nitrogen removal.

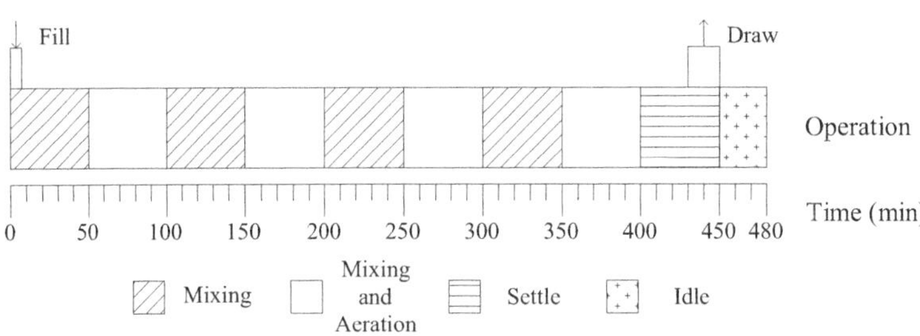

Figure 6.9 An example of intermittently aerated sequencing batch reactors (Li *et al.* 2008).

Wang *et al.* (2016) examined N_2O emission in a multiple anoxic and aerobic process at the aeration rates of 600 mL/min (SBR_L) and 1200 mL/min (SBR_H). The nitrogen removal was high, 89%, in SBR_L in comparison with SBR_H, 71%. N_2O

emission mainly occurred during the aerobic phase, and the N_2O emission factor was 10.1% in SBR_L and 2.3% in SBR_H, much higher in SBR_L than in SBR_H. The low aeration condition enhanced N_2O generation within aeration phases due to the low DO concentrations, impacting nitrification. Zhang *et al.* (2006) investigated N_2O emission in a two-step feeding multiple A/O process and found that N_2O emission mainly occurred under the second anoxic and aerobic phase, because the low DO, nitrite and organic carbon concentrations enhanced N_2O emission through both nitrification and denitrification processes. De Mello *et al.* (2013) observed that in a full-scale intermittently aerated WWTP, the N_2O emission factor was 0.1% of the influent TN loading rate and its emission mainly occurred during the aeration stage.

Intermittently aerated sequencing batch reactor (IASBR) technology can create a suitable condition for NO_2^- accumulation (i.e., partial nitrification) (Li *et al.* 2011). However, due to the accumulation of NO_2^-, a high potential of N_2O emission has often been observed. In an IASBR process, the N_2O emission factor was 15.6% when treating pig manure digestate and was 10.1% when treating synthetic wastewater made of easily biodegradable organic matter (Zhang *et al.* 2012). N_2O emission also mainly occurred during the aerobic phase, up to 92% of the total. When treating pig manure digestate, in an IASBR operation cycle, NO_2^--N concentrations were up to 650–720 mg/L but only 40–72 mg/L when treating synthetic wastewater. The lower N_2O emission when treating synthetic wastewater was because of the rapid consumption of NO_2^- by denitrifiers with the easily biodegradable matter in the non-aeration phase.

Osada *et al.* (1995) used the intermittent aeration operating mode to decrease the N_2O emission factor from 35% to 1% of the nitrogen load, possibly due to the avoidance of NO_2^- accumulation. Kimochi *et al.* (1998) found that in a full-scale intermittently aerated WWTP, N_2O emission was 0.43–1.89 g/person/year, and its emission factor was 0.03–0.11%.

By adjusting the anoxic to aerobic phase ratio, N_2O emission may be reduced. Itokawa *et al.* (1996) found that N_2O emissions in an intermittent aeration tank (IAT) were in the range of 0.24–55%, and the emission was mainly due to incomplete nitrification and denitrification, which could be reduced by adjusting the anoxic to aerobic phase ratio and by dosing methanol. When treating pig slurry at a TN loading rate of 610 mg/L/day with a removal percentage of 52–54%, the N_2O emission factor was 18% at an aerobic to anoxic ratio of 0.625, and the N_2O emission was almost zero at the ratio of 0.375, leading to complete denitrification (Beline & Martinez, 2002).

By adopting intermittent aeration, energy consumption in WWTPs can be reduced, while this may increase the N_2O emission factor from 0.07% to 27%, enhance the growth of filamentous microorganisms and increase the effluent turbidity (Dotro *et al.* 2011). Future studies may consider the integrated activated sludge and biofilm system for reducing N_2O emission and enhancing nitrogen removal. Park *et al.* (2000) found that in the intermittently aerated system (aerobic phase of 60 min and anoxic phase of 30 min), the N_2O emission factor was 4.57% in the suspended activated sludge system and 3% in the biofilm system, and the emission was reduced to below 0.2% with the dosage of methanol for complete denitrification; biofilm systems might enhance the diversity of microbial communities and benefit the reduction of N_2O emission.

6.5.3 Denitrifying polyphosphate accumulating processes

The enhanced biological phosphorus removal process consists of sequential anaerobic and aerobic/anoxic phases (like the A^2O process mentioned above). Under anoxic conditions, some polyphosphate accumulating organisms (PAOs) can use stored organic carbon (such as PHB) as the substrate for denitrification. This process is a denitrifying polyphosphate accumulating process and these PAOs are denitrifying PAOs, i.e., DNPAOs. The degradation of intracellular organic carbons is a rate limiting step, which may induce electron donor competition and further enhance a high N_2O emission. The possible reasons for this include (i) the intrinsic PHB utilization, and/or (ii) NO_2^- accumulation, especially in granular systems.

In the study of Tang *et al.* (2016), DNPAOs were enriched in a lab-scale reactor. As shown in Figure 6.10, during denitrifying NO_3^-, a very low amount of N_2O was produced, independent of the applied different NO_3^- concentrations. However, with NO_2^- as the electron acceptor, a much higher N_2O emission occurred. The N_2O emission ratio to the denitrified NO_2-N was 6.2%, 5.3% and 4.9% at the initial NO_2-N concentration of 10, 20 and 40 mg/L, respectively. When stored organic carbon (PHA) in DNPAOs was used as the electron donor, N_2O emission was dependent on the electron acceptor. When NO_3^- was the electron acceptor, there was not much N_2O emission, while when NO_2^- was the electron acceptor, a high N_2O emission was observed. Therefore, PHA was not the intrinsic factor causing N_2O emission during denitrification for the acclimated DNPAOs but the electron acceptor played an important role.

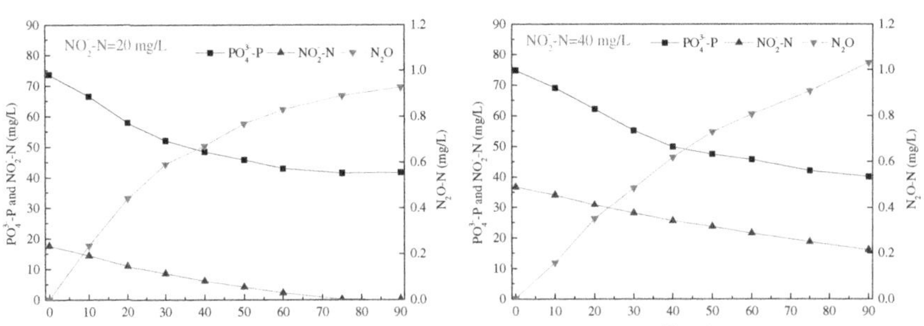

Figure 6.10 N_2O emission under different electron acceptor concentrations by DNPAOs (Tang *et al.* 2016).

The denitrifying polyphosphate accumulating process is found to have high N_2O emission factors. In a simultaneous nitrogen and phosphorus removal system, the N_2O emission factor was around 75% of the removed ammonia, which was due to activities of glycogen accumulating organisms (GAOs) rather than PAOs, and this only occurred with NO_2-N above 1 mg/L (Zeng *et al.* 2003). In a simultaneous nitrification, denitrification and phosphorus removal process, with a low applied DO concentration, the N_2O emission factor was around 51% due to denitrification by GAOs (with intracellular organic carbon as the electron donor), and this factor was increased to 77% with the addition of NO_2^- (Lemaire *et al.* 2006). Meyer *et al.* (2005) noticed that in a simultaneous nitrification, denitrification and phosphorus removal process, N_2O was the main product of denitrification with PHA as the organic carbon combined with even a low concentration of nitrite.

In denitrifying polyphosphate accumulating processes, research has found that for even a low concentration of NO_2^-, above 1 mg/L, N_2O would be produced. This does not happen with NO_3^-.

Wang *et al.* (2011b) found that N_2O emission from a denitrifying phosphorus removal system was 7.77% of the removed TN from wastewater, while it was 13.93%, 37.95% and 24.71% when the organic carbon sources were acetate, mixture of acetate and propionate, and propionate, respectively. The high N_2O emission was due to (1) decreased denitrifying PAO (DNPAO) activities and accumulated NO_2^-, (2) the high proportion of polyhydroxyvalerate (PHV) which was degraded slowly, and (3) the oxidative stress of producing N_2O and enhancing nitrifier denitrification. Li *et al.* (2013a) examined shock loading on N_2O emission from a denitrifying phosphorus removal system, and when COD was increased from 200 to 350 or 500 mg/L, the N_2O emission factor increased from 1.62% to 7.12% or 3.29%, respectively; while when COD was decreased to 100 mg/L, the factor was 1.20%. Under high influent COD conditions, more PHA was produced and induced a high NO_2^- accumulation, resulting in a high N_2O emission.

Zhou *et al.* (2008) found that FNA rather than NO_2^- inhibited DNPAOs, with 50% inhibition at FNA of 0.0007–0.001 mg/L and complete inhibition at 0.004 mg/L. The inhibition was reversible and dependent on the inhibition concentration. The inhibition leads to N_2O emission.

6.5.4 High ammonium wastewater treatment processes

For wastewater with a high NH_4^+ concentration and a low C/N ratio, partial nitrification (i.e., nitritation) followed by denitrification or anaerobic ammonia oxidization (Anammox) processes have been investigated and applied to treat this type of wastewater. In these processes, NO_2^- accumulation through partial nitrification is a prerequisite, which may induce a high N_2O emission.

6.5.4.1 *Partial nitrification*

For high NH_4^+-containing wastewater, the intention for partial nitrification generates more N_2O emission than complete denitrification because of the high concentration of NO_2^- in the nitrifying tanks. Schneider *et al.* (2013) found that the N_2O emission factor was 2.9% for partial nitrification, and 0.74% for complete nitrification. When switching from complete nitrification to partial nitrification, a high N_2O emission factor of 4.81% was observed. All these studies indicate that high N_2O emission from partial nitrification facilities is intrinsic and not avoidable. Hence, complete oxidation of NH_4^+ to NO_3^- with sufficient aeration duration would be a minimization strategy for N_2O emission control.

6.5.4.2 *Partial nitrification-Anammox process*

In the Anammox process, Anammox bacteria use NO_2^- as the preferred electron acceptor to oxidize NH_4^+ (Eq. 6.5), and use bicarbonate as the sole carbon source.

$$\begin{aligned} &NH_4^+ + 1.32NO_2^- + 0.066HCO_3^- + 0.13H^+ \\ &\quad \rightarrow 0.066CH_2O_{0.5}N_{0.15} + 1.02N_2 + 0.26NO_3^- + 2.03H_2O \end{aligned} \tag{6.5}$$

For the partial nitrification–Anammox process (PNA), partial nitrification and Anammox may occur within one reactor (like CANON – completely autotrophic nitrogen removal over nitrite – process) or in two reactors: one reactor for achieving oxidation of NH_4^+ to NO_2^- (i.e., partial nitrification) and the other reactor for achieving nitrogen removal through Anammox. The two-reactor PNA requires a high investment but has a high reaction rate. In addition, the first reactor can remove COD, which has been found to inhibit Anammox activity.

N_2O generated from PNA is most likely produced by ammonium oxidizers in the partial nitrification (De Graaff *et al.* 2010), so partial nitrification rather than Anammox is the main source of N_2O emission. The two main N_2O emission pathways are the oxidation of hydroxylamine (NH_2OH) during ammonia oxidation, and the reduction of NO_2^- to N_2O (nitrifier denitrification) as described in Chapter 1. Anammox organisms don't reduce NO_3^- or NO_2^- through conventional denitrification via N_2O, but they may be the indirect source of N_2O by producing NO, which can be used by AOB for producing N_2O (Kartal *et al.* 2007; Weissenbacher *et al.* 2011). Okabe *et al.* (2011) found that the N_2O emission factors in the partial nitrification and Anammox reactors, were 4% and 0.1%, respectively, of the influent nitrogen load. In a single reactor PNA process, since NO_2^- converted from NH_4^+ is consumed by Anammox bacteria, its concentrations are kept relatively low. As a consequence, N_2O emission from the one-reactor PNA systems is in the range of 0.4–1.3%, while it is 1.7–6.6% in the two-reactor PNA systems (Kampschreur *et al.* 2008; Joss *et al.* 2009; Kampschreur *et al.* 2009b; Weissenbacher *et al.* 2010; Desloover *et al.* 2011).

Apart from NO_2^-, in PNA systems, other factors including DO concentration, NH_4^+ load, granule size, and pH have been found to influence N_2O production. Increasing oxygen, decreasing pH, decreasing temperature (25–40°C) and high concentrations of nitrogen in the influent wastewater increase accumulation of N_2O (Shao *et al.* 2011; Van Hulle *et al.* 2012).

Granule size

Granule sizes need to be large enough to not limit optimal Anammox activity, but larger particles (granule size: 0.5 mm, 0.75 mm, and 1 mm) could cause an elevated N_2O production (Van Hulle *et al.* 2012). Large granule particles would encourage heterotrophic denitrification inside the granules, leading to N_2O production.

DO

N_2O emissions increase with decreasing DO concentrations in a partial nitrification reactor of a full-scale two-reactor PNA system (Kampschreur *et al.* 2008). However, in single reactor PNA systems, intense aeration may result in higher N_2O emission and formation than during low aeration periods, with the mean N_2O formation rate of 0.050 kg N/m^3/day at high aeration of 4041 m^3/h and 0.029 kg N/m^3/day at low aeration of 1088 m^3/h in a full-scale one-reactor PNA system (Castro-Barros *et al.* 2015). This is due to the inhibition of DO on Anammox activity.

pH

N_2O emissions are strongly dependent on pH which determines FNA concentration in reactors. Similar to heterotrophic denitrification, N_2O emissions from PNA processes increase as pH decreases (Shiskowski & Mavinic, 2006; Rathnayake *et al.* 2015). For example, the N_2O production at a pH of 6.5 is about 11 times higher than that at a pH of 8.0 in Anammox granules (Okabe *et al.* 2011).

N concentrations in the influent wastewater

In the PNA process, a higher conversion rate of NH_4^+ to N_2O occurs when a higher concentration of NH_4^+-N is applied in the wastewater (Van Hulle *et al.* 2012). For instance, in a single reactor PNA, *in situ* N_2O production was low with less than 0.05 mg N_2O-N/L under low NH_4^+condition (<50 mg/L), whereas high N_2O productions with around 0.15 mg N_2O-N/L were observed under high NH_4^+condition (150 mg/L) (Ali *et al.* 2016).

6.5.5 N_2O emission in autotrophic denitrification process

Biological autotropic denitrification is an anaerobic process facilitated by autotrophic denitrifiers which can utilize a variety of electron donors, including ammonium (namely the Anammox process discussed above), hydrogen (Mao *et al.* 2013), and reduced sulfur compounds (Tang *et al.* 2010; Li *et al.* 2016), with oxidized nitrogen (NO_3^- and NO_2^-) as the electron acceptor. Denitrification with element S, FeS_2 and FeS can proceed in accordance with the stoichiometric equations in Eq. 6.6, 6.7 and 6.8 (Juncher Jørgensen *et al.* 2009; Yang *et al.* 2017a; Yang *et al.* 2017b):

$$S + 1.2NO_3^- + 0.4H_2O \rightarrow SO_4^{2-} + 0.6N_2 + 0.8H^+ \quad (6.6)$$

$$5FeS_2 + 14NO_3^- + 2H_2O \rightarrow 5Fe^{2+} + 10SO_4^{2-} + 4OH^- + 7N_2 \quad (6.7)$$

$$5FeS + 8NO_3^- + 4H_2O \rightarrow 4N_2 + 5SO_4^{2-} + 5Fe^{2+} + 8OH^- \quad (6.8)$$

Denitrification with H_2 as the electron donor is described as:

$$NO_3^- + H_2 \rightarrow NO_2^- + H_2O \quad (6.9)$$

$$2NO_2^- + 3H_2 \rightarrow N_2 + 2OH^- + 2H_2O \quad (6.10)$$

Very few studies have been conducted to investigate N_2O emissions from H_2 and sulfur based autotrophic denitrification systems.

In the elemental sulfur autotrophic denitrification processes, the amount of N_2O production per unit mass of nitrogen with NO_2^- as the electron acceptor is greater than with NO_3^- as the electron acceptor, and N_2O production is directly related to the concentration of NO_2^- and NO_3^- (Park *et al.* 2002). Meanwhile no N_2O was observed in the lab-scale experiment at the NO_3^- concentrations ranging from 124 to 2480 mg/L in sulfide-based autotrophic denitrification (Shao *et al.* 2011).

Sulfide stimulates N_2O reduction in sulfide-driven autotrophic denitrification (Yang *et al.* 2016). For sulfide-driven autotrophic denitrifying granular sludge (granule size: 701 mm), at S/N of 5.0 g S/g N, N_2O reduction rate was 27.7 mg N/g VSS/h; this was 2 and 4 times greater than those at 2.5 and 0.8 g S/g N, respectively. The N_2O emission factors from the reactor significantly increased from 0.002% of the N load to 0.405% of the N load as the S/N mass ratio in the wastewater decreased from 2.1 to 1.4 g S/g N (Yang *et al.* 2016). Limitation of sulfide leads to N_2O accumulation.

Table 6.1 summarizes N_2O emissions from autotrophic denitrification processes. The low N_2O emission from sulfur-based autotrophic denitrification process reduces potential risks of a significant amount of N_2O emissions, which is very small compared with heterotrophic denitrification (Yang *et al.* 2016).

6.6 SUMMARY AND FUTURE TRENDS

Nitrogen removal is associated with various metabolic functions, which are carried out by different types of microbial communities. Therefore, N_2O generation during nitrification and denitrification is complicated. Factors influencing N_2O

Table 6.1 N_2O emission from autotrophic denitrification processes.

Processes	N_2O Emission (% of N-load)	N_2O Emission (% of N-load)	Wastewater	N Concentration or Loadings	References
Laboratory-scale two-reactor PNA	4.0 ± 1.5% (partial nitrification)	0.1 ± 0.07% (Annammox)	Artificial wastewater	300 mg NH_4^+-N/L	Okabe *et al.* (2011)
Full-scale two-reactor PNA	1.7% (partial nitrification)	0.6% (Annammox)	Reject water (the centrate of the digested sludge)	1200–1600 mg NH_4^+-N/L	Kampschreur *et al.* (2008)
Full-scale single-stage PNA	1.2%		Wastewater from a potato processing factory and reject water of a municipal sludge dewatering plant	250 mg NH_4^+-N/L	Kampschreur *et al.* (2009a)
Full-scale single-reactor PNA	0.63% of eliminated N		Supernatant taken from the digester of WWTPs	600–990 mg NH_4^+-N/L	Joss *et al.* (2009)
Laboratory-scale single-stage PNA	<0.1%		Synthetic wastewater	66–128 mg NH_4^+-N/L	Sliekers *et al.* (2002)
Full-scale one-stage PNA	6.28%		Wastewater from a potato processing factory and reject water of a municipal sludge dewatering plant	1.75 kg NH_4^+-N/m^3/day	Castro-Barros *et al.* (2015)
Sulfur denitrification	0.04–0.84% of eliminated N		Real groundwater	45 mg NO_3^--N/L	Hasegawa *et al.* (2004)
Sulfur denitrification	0.3–0.5% of eliminated N		Synthetic groundwater	25–80 mg NO_3^--N/L	Zhang *et al.* (2015)

generation include the diverse microbial communities, environmental factors and operating conditions. Further studies should explore advanced techniques such as isotopic analysis, molecular techniques and the concept of microbial community competition in studying N_2O generation mechanisms.

Understanding N_2O generation mechanisms and influencing factors is important for the purpose of N_2O emission mitigation. N_2O emission should be mitigated through emission source control and end treatment. The emission source control is the focus at present, while the end treatment should be focused on in future. Source control strategies include minimizing N_2O generation and emission during aerobic nitrification and anoxic denitrification, and maximizing N_2O consumption under anoxic conditions (Desloover *et al.* 2012). The practical strategies include: (1) to provide sufficient organic carbon for heterotrophic denitrifiers by adding external carbon sources to wastewater with low C/N ratio. Addition of methanol into a BNR process reduced N_2O emission from 4.5% to 0.2% of the nitrogen load (Park *et al.* 2000); (2) to avoid high NO_2^- accumulation in the nitrifying tanks. The practices are: avoidance of NO_2^- accumulation in partial nitrification systems and achievement of complete oxidation of NH_4^+ (or most likely hydroxylamine) as a result of sufficient aeration time; and in-lane consumption of NO_2^- produced with simultaneous nitrification and denitrification process, one reactor PNA process and well controlled intermittent aeration process (Weissenbacher *et al.* 2011); and (3) to maximize reduction of N_2O that is produced in preceding water tanks under anoxic conditions.

6.7 REFERENCES

Aboobakar A., Cartmell E., Stephenson T., Jones M., Vale P. and Dotro G. (2013). Nitrous oxide emissions and dissolved oxygen profiling in a full-scale nitrifying activated sludge treatment plant. *Water Research*, **47**(2), 524–534.

Ali M., Rathnayake R. M., Zhang L., Ishii S., Kindaichi T., Satoh H., Toyoda S., Yoshida N. and Okabe S. (2016). Source identification of nitrous oxide emission pathways from a single-stage nitritation-anammox granular reactor. *Water Research*, **102**, 147–157.

Alinsafi A., Adouani N., Béline F., Lendormi T., Limousy L. and Sire O. (2008). Nitrite effect on nitrous oxide emission from denitrifying activated sludge. *Process Biochemistry*, **43**(6), 683–689.

Beline F. and Martinez J. (2002). Nitrogen transformations during biological aerobic treatment of pig slurry: effect of intermittent aeration on nitrous oxide emissions. *Bioresource Technology*, **83**(3), 225–228.

Beline F., Martinez J., Marol C. and Guiraud G. (2001). Application of the ^{15}N technique to determine the contributions of nitrification and denitrification to the flux of nitrous oxide from aerated pig slurry. *Water Research*, **35**(11), 2774–2778.

Belmonte M., Vázquez-Padín J., Figueroa M., Campos J., Méndez R., Vidal G. and Mosquera-Corral A. (2012). Denitrifying activity via nitrite and N_2O production using acetate and swine wastewater. *Process Biochemistry*, **47**(7), 1202–1206.

Bonin P., Gilewicz M. and Bertrand J. (1989). Effects of oxygen on each step of denitrification on *Pseudomonas nautica*. *Canadian Journal of Microbiology*, **35**(11), 1061–1064.

Castro-Barros C. M., Daelman M. R., Mampaey K. E., van Loosdrecht M. C. and Volcke E. I. (2015). Effect of aeration regime on N_2O emission from partial nitritation-anammox in a full-scale granular sludge reactor. *Water Research*, **68**, 793–803.

Chen Y., Wang D., Zhu X., Zheng X. and Feng L. (2012). Long-term effects of copper nanoparticles on wastewater biological nutrient removal and N_2O generation in the activated sludge process. *Advances in Environmental Science and Technology*, **46**(22), 12452–12458.

Chen Y., Wang D., Zheng X., Li X., Feng L. and Chen H. (2014). Biological nutrient removal with low nitrous oxide generation by cancelling the anaerobic phase and extending the idle phase in a sequencing batch reactor. *Chemosphere*, **109**, 56–63.

Chung Y.-C. and Chung M.-S. (2000). BNP test to evaluate the influence of C/N ratio on N_2O production in biological denitrification. *Water Science and Technology*, **42**(3–4), 23–27.

De Graaff M., Zeeman G., Temmink H., van Loosdrecht M. and Buisman C. (2010). Long term partial nitritation of anaerobically treated black water and the emission of nitrous oxide. *Water Research*, **44**(7), 2171–2178.

Desloover J., De Clippeleir H., Boeckx P., Du Laing G., Colsen J., Verstraete W. and Vlaeminck S. E. (2011). Floc-based sequential partial nitritation and anammox at full scale with contrasting N_2O emissions. *Water Research*, **45**(9), 2811–2821.

Desloover J., Vlaeminck S. E., Clauwaert P., Verstraete W. and Boon N. (2012). Strategies to mitigate N_2O emissions from biological nitrogen removal systems. *Current Opinion in Biotechnology*, **23**(3), 474–482.

Dotro G., Jefferson B., Jones M., Vale P., Cartmell E. and Stephenson T. (2011). A review of the impact and potential of intermittent aeration on continuous flow nitrifying activated sludge. *Environmental Technology*, **32**(15), 1685–1697.

Felgate H., Giannopoulos G., Sullivan M. J., Gates A. J., Clarke T. A., Baggs E., Rowley G. and Richardson D. J. (2012). The impact of copper, nitrate and carbon status on the emission of nitrous oxide by two species of bacteria with biochemically distinct denitrification pathways. *Applied and Environmental Microbiology*, **14**(7), 1788–1800.

Foley J., Yuan Z., Keller J., Senante E., Chandran K., Willis J., Shah A., van Loosdrecht M. and van Voorthuizen E. (2011). N_2O and CH_4 Emission from Wastewater Collection and Treatment Systems: Technical Report. Global Water Research Coalition, London, United Kingdom.

Gejlsbjerg B., Frette L. and Westermann P. (1998). Dynamics of N_2O production from activated sludge. *Water Research*, **32**(7), 2113–2121.

Glass J. B. and Orphan V. J. (2012). Trace metal requirements for microbial enzymes involved in the production and consumption of methane and nitrous oxide. *Frontiers in Microbiology*, **3**.

Granger J. and Ward B. B. (2003). Accumulation of nitrogen oxides in copper-limited cultures of denitrifying bacteria. *Limnology and Oceanography*, **48**(1), 313–318.

Groffman P. M., Altabet M. A., Böhlke J., Butterbach-Bahl K., David M. B., Firestone M. K., Giblin A. E., Kana T. M., Nielsen L. P. and Voytek M. A. (2006). Methods for measuring denitrification: diverse approaches to a difficult problem. *Ecological Applications*, **16**(6), 2091–2122.

Hanaki K., Hong Z. and Matsuo T. (1992). Production of nitrous oxide gas during denitrification of wastewater. *Water Science and Technology*, **26**(5–6), 1027–1036.

Hasegawa K., Shimizu K. and Hanaki K. (2004). Nitrate removal with low N_2O emission by application of sulfur denitrification in actual agricultural field. *Water Science and Technology*, **50**(8), 145–151.

Hu Z., Zhang J., Li S., Xie H., Wang J., Zhang T., Li Y. and Zhang H. (2010). Effect of aeration rate on the emission of N_2O in anoxic–aerobic sequencing batch reactors (A/O SBRs). *Journal of Bioscience and Bioengineering*, **109**(5), 487–491.

Hu Z., Zhang J., Li S., Wang J. and Zhang T. (2011). Effect of anoxic/aerobic phase fraction on N_2O emission in a sequencing batch reactor under low temperature. *Bioresource Technology*, **102**(9), 5486–5491.

Ito O. and Matsumura M. (2001). Effect of copper sulfate addition on nitrous-oxide emission in high speed denitrification process using macro-porous cellulose carrier. *Environmental Technology*, **22**(3), 345–353.

Itokawa H., Hanaki K. and Matsuo T. (1996). Nitrous oxide emission during nitrification and denitrification in a full-scale night soil treatment plant. *Water Science and Technology*, **34**(1–2), 277–284.

Jian M., Jiang X.-Q., Lin-Zhang Y., Zhang J., Qing-Yun Q., Chen-Da H. and Shi-Xue Y. (2006). Nitrous oxide production in a sequence batch reactor wastewater treatment system using synthetic wastewater. *Pedosphere*, **16**(4), 451–456.

Joss A., Salzgeber D., Eugster J., König R., Rottermann K., Burger S., Fabijan P., Leumann S., Mohn J. and Siegrist H. (2009). Full-scale nitrogen removal from digester liquid with partial nitritation and anammox in one SBR. *Environmental Science & Technology*, **43**(14), 5301–5306.

Juncher Jørgensen C., Jacobsen O. S., Elberling B. and Aamand J. (2009). Microbial oxidation of pyrite coupled to nitrate reduction in anoxic groundwater sediment. *Environmental Science & Technology*, **43**(13), 4851–4857.

Kampschreur M. J., van der Star W. R., Wielders H. A., Mulder J. W., Jetten M. S. and van Loosdrecht M. C. (2008). Dynamics of nitric oxide and nitrous oxide emission during full-scale reject water treatment. *Water Research*, **42**(3), 812–826.

Kampschreur M., Poldermans R., Kleerebezem R., VanDer Star W., Haarhuis R., Abma W., Jetten M. and Van Loosdrecht M. (2009a). Emission of nitrous oxide and nitric oxide from a full-scale single-stage nitritation-anammox reactor. *Water Science and Technology*, **60**(12), 3211–3217.

Kampschreur M. J., Temmink H., Kleerebezem R., Jetten M. S. and van Loosdrecht M. C. (2009b). Nitrous oxide emission during wastewater treatment. *Water Research*, **43**(17), 4093–4103.

Kartal B., Rattray J., van Niftrik L. A., van de Vossenberg J., Schmid M. C., Webb R. I., Schouten S., Fuerst J. A., Damsté J. S. and Jetten M. S. (2007). Candidatus "*Anammoxoglobus propionicus*" a new propionate oxidizing species of anaerobic ammonium oxidizing bacteria. *Systematic and Applied Microbiology*, **30**(1), 39–49.

Kim D. J. and Kim Y. R. (2011). Effect of ammonium concentration on the emission of N_2O under oxygen-limited autotrophic wastewater nitrification. *Journal of Microbiology and Biotechnology*, **21**(9), 988–994.

Kim S.-W., Miyahara M., Fushinobu S., Wakagi T. and Shoun H. (2010). Nitrous oxide emission from nitrifying activated sludge dependent on denitrification by ammonia-oxidizing bacteria. *Bioresource Technology*, **101**(11), 3958–3963.

Kimochi Y., Inamori Y., Mizuochi M., Xu K.-Q. and Matsumura M. (1998). Nitrogen removal and N_2O emission in a full-scale domestic wastewater treatment plant with intermittent aeration. *Journal of Fermentation and Bioengineering*, **86**(2), 202–206.

Kishida N., Kim J., Kimochi Y., Nishimura O., Sasaki H. and Sudo R. (2004). Effect of C/N ratio on nitrous oxide emission from swine wastewater treatment process. *Water Science and Technology*, **49**(5–6), 359–371.

Law Y., Lant P. and Yuan Z. (2011). The effect of pH on N_2O production under aerobic conditions in a partial nitritation system. *Water Research*, **45**(18), 5934–5944.

Lemaire R., Meyer R., Taske A., Crocetti G. R., Keller J. and Yuan Z. (2006). Identifying causes for N_2O accumulation in a lab-scale sequencing batch reactor performing simultaneous nitrification, denitrification and phosphorus removal. *EJB Electronic Journal of Biotechnology*, **122**(1), 62–72.

Li C., Wang T., Zheng N., Zhang J., Ngo H. H., Guo W. and Liang S. (2013). Influence of organic shock loads on the production of N_2O in denitrifying phosphorus removal process. *Bioresource Technology*, **141**, 160–166.

Li J., Elliott D., Nielsen M., Healy M. G. and Zhan X. (2011). Long-term partial nitrification in an intermittently aerated sequencing batch reactor (SBR) treating ammonium-rich wastewater under controlled oxygen-limited conditions. *Biochemical Engineering Journal*, **55**(3), 215–222.

Li Q., Li P., Zhu P., Wu J. and Liang S. (2008). Effects of exogenous organic carbon substrates on nitrous oxide emissions during the denitrification process of sequence batch reactors. *Environmental Engineering Science*, **25**(8), 1221–1228.

Li R., Morrison L., Collins G., Li A. and Zhan X. (2016). Simultaneous nitrate and phosphate removal from wastewater lacking organic matter through microbial oxidation of pyrrhotite coupled to nitrate reduction. *Water Research*, **96**, 32–41.

Lim Y. and Kim D.-J. (2014). Quantification method of N_2O emission from full-scale biological nutrient removal wastewater treatment plant by laboratory batch reactor analysis. *Bioresource Technology*, **165**, 111–115.

Lu H. and Chandran K. (2010). Factors promoting emissions of nitrous oxide and nitric oxide from denitrifying sequencing batch reactors operated with methanol and ethanol as electron donors. *Biotechnology and Bioengineering*, **106**(3), 390–398.

Mao Y., Xia Y. and Zhang T. (2013). Characterization of Thauera-dominated hydrogen-oxidizing autotrophic denitrifying microbial communities by using high-throughput sequencing. *Bioresource Technology*, **128**, 703–710.

Mello W. Z. d., Ribeiro R. P., Brotto A. C., Kligerman D. C., Piccoli A. d. S. and Oliveira J. L. (2013). Nitrous oxide emissions from an intermittent aeration activated sludge system of an urban wastewater treatment plant. *Quim Nova*, **36**(1), 16–20.

Meyer R. L., Zeng R. J., Giugliano V. and Blackall L. L. (2005). Challenges for simultaneous nitrification, denitrification, and phosphorus removal in microbial aggregates: mass transfer limitation and nitrous oxide production. *FEMS Microbiology Ecology*, **52**(3), 329–338.

Ni B.-J., Peng L., Law Y., Guo J. and Yuan Z. (2014). Modeling of nitrous oxide production by autotrophic ammonia-oxidizing bacteria with multiple production pathways. *Advances in Environmental Science and Technology*, **48**(7), 3916–3924.

Noda N., Kaneko N., Mikami M., Kimochi Y., Tsuneda S., Hirata A., Mizuochi M. and Inamori Y. (2004). Effects of SRT and DO on N_2O reductase activity in an anoxic-oxic activated sludge system. *Water Science and Technology*, **48**(11–12), 363–370.

Okabe S., Oshiki M., Takahashi Y. and Satoh H. (2011). N_2O emission from a partial nitrification–anammox process and identification of a key biological process of N_2O emission from anammox granules. *Water Research*, **45**(19), 6461–6470.

Osada T., Kuroda K. and Yonaga M. (1995). Reducing nitrous oxide gas emissions from fill-and-draw type activated sludge process. *Water Research*, **29**(6), 1607–1608.

Pan Y., Ye L., Ni B.-J. and Yuan Z. (2012). Effect of pH on N_2O reduction and accumulation during denitrification by methanol utilizing denitrifiers. *Water Research*, **46**(15), 4832–4840.

Pan Y., Ni B.-J., Bond P. L., Ye L. and Yuan Z. (2013). Electron competition among nitrogen oxides reduction during methanol-utilizing denitrification in wastewater treatment. *Water Research*, **47**(10), 3273–3281.

Park J.-H., Shin H.-S., Lee I.-S. and Bae J.-H. (2002). Denitrification of high NO_3^--N containing wastewater using elemental sulfur; nitrogen loading rate and N_2O production. *Environmental Technology*, **23**(1), 53–65.

Park K. Y., Inamori Y., Mizuochi M. and Ahn K. H. (2000). Emission and control of nitrous oxide from a biological wastewater treatment system with intermittent aeration. *Journal of Bioscience and Bioengineering*, **90**(3), 247–252.

Peng L., Ni B.-J., Erler D., Ye L. and Yuan Z. (2014). The effect of dissolved oxygen on N_2O production by ammonia-oxidizing bacteria in an enriched nitrifying sludge. *Water Research*, **66**, 12–21.

Perez-Garcia O., Villas-Boas S. G., Swift S., Chandran K. and Singhal N. (2014). Clarifying the regulation of NO/N_2O production in Nitrosomonas europaea during anoxic–oxic transition via flux balance analysis of a metabolic network model. *Water Research*, **60**, 267–277.

Poth M. and Focht D. D. (1985). ^{15}N kinetic analysis of N_2O production by *Nitrosomonas europaea*: an examination of nitrifier denitrification. *Applied and Environmental Microbiology*, **49**(5), 1134–1141.

Rassamee V., Sattayatewa C., Pagilla K. and Chandran K. (2011). Effect of oxic and anoxic conditions on nitrous oxide emissions from nitrification and denitrification processes. *Biotechnology and Bioengineering*, **108**(9), 2036–2045.

Rathnayake R. M., Oshiki M., Ishii S., Segawa T., Satoh H. and Okabe S. (2015). Effects of dissolved oxygen and pH on nitrous oxide production rates in autotrophic partial nitrification granules. *Bioresource Technology*, **197**, 15–22.

Remde A. and Conrad R. (1990). Production of nitric oxide in Nitrosomonas europaea by reduction of nitrite. *Archives of Microbiology*, **154**(2), 187–191.

Richardson D., Felgate H., Watmough N., Thomson A. and Baggs E. (2009). Mitigating release of the potent greenhouse gas N_2O from the nitrogen cycle–could enzymic regulation hold the key? *Trends in Biotechnology*, **27**(7), 388–397.

Schalk-Otte S., Seviour R. J., Kuenen J. and Jetten M. (2000). Nitrous oxide (N_2O) production by Alcaligenes faecalis during feast and famine regimes. *Water Research*, **34**(7), 2080–2088.

Schneider Y., Beier M. and Rosenwinkel K.-H. (2013). Nitrous oxide formation during nitritation and nitrification of high-strength wastewater. *Water Science and Technology*, **67**(11), 2494–2502.

Schreiber F., Wunderlin P., Udert K. M. and Wells G. F. (2012). Nitric oxide and nitrous oxide turnover in natural and engineered microbial communities: biological pathways, chemical reactions, and novel technologies. *Frontiers in Microbiology*, **3**.

Shao M.-F., Zhang T., Fang H. H.-P. and Li X. (2011). The effect of nitrate concentration on sulfide-driven autotrophic denitrification in marine sediment. *Chemosphere*, **83**, 1–6.

Shiskowski D. M. and Mavinic D. S. (2006). The influence of nitrite and pH (nitrous acid) on aerobic-phase, autotrophic N2O generation in a wastewater treatment bioreactor. *Journal of Environmental Engineering and Science*, **5**(4), 273–283.

Sliekers A. O., Derwort N., Gomez J. C., Strous M., Kuenen J. and Jetten M. (2002). Completely autotrophic nitrogen removal over nitrite in one single reactor. *Australian Journal of Marine and Freshwater Research*, **36**(10), 2475–2482.

Sun S., Cheng X., Li S., Qi F., Liu Y. and Sun D. (2013). N_2O emission from full-scale urban wastewater treatment plants: a comparison between A^2O and SBR. *Water Science and Technology*, **67**(9), 1887–1893.

Takaya N., Catalan-Sakairi M. A. B., Sakaguchi Y., Kato I., Zhou Z. and Shoun H. (2003). Aerobic denitrifying bacteria that produce low levels of nitrous oxide. *Applied and Environmental Microbiology*, **69**(6), 3152–3157.

Tallec G., Garnier J., Billen G. and Gousailles M. (2006). Nitrous oxide emissions from secondary activated sludge in nitrifying conditions of urban wastewater treatment plants: effect of oxygenation level. *Water Research*, **40**(15), 2972–2980.

Tang B., Wu G., Wang H. and Guan Y. (2016). Nitrous oxide emission depending on the type of electron acceptor by a denitrifying phosphorus removal sludge. *Global NEST Journal*, **18**(2), 251–258.

Tang S., Yang Q., Shang H. and Sun T. (2010). Removal of nitrate by autosulfurotrophic denitrifying bacteria: optimization, kinetics and thermodynamics study. *Fresenius Environmental Bulletin*, **19**, 3193–3198.

Thörn M. and Sörensson F. (1996). Variation of nitrous oxide formation in the denitrification basin in a wastewater treatment plant with nitrogen removal. *Water Research*, **30**(6), 1543–1547.

Van Hulle S. W., Callens J., Mampaey K. E., van Loosdrecht M. C. and Volcke E. I. (2012). N_2O and NO emissions during autotrophic nitrogen removal in a granular sludge reactor – a simulation study. *Environmental Technology*, **33**(19–21), 2281–2290.

Von Schulthess R., Wild D. and Gujer W. (1994). Nitric and nitrous oxides from denitrifying activated sludge at low oxygen concentration. *Water Science and Technology*, **30**(6), 123–132.

Wang H., Guan Y., Pan M. and Wu G. (2016). Aerobic N_2O emission for activated sludge acclimated under different aeration rates in the multiple anoxic and aerobic process. *Journal of Environmental Sciences*, **43**, 70–79.

Wang Y., Geng J., Guo G., Wang C. and Liu S. (2011). N_2O production in anaerobic/anoxic denitrifying phosphorus removal process: the effects of carbon sources shock. *Chemical Engineering Journal*, **172**(2), 999–1007.

Weissenbacher N., Takacs I., Murthy S., Fuerhacker M. and Wett B. (2010). Gaseous nitrogen and carbon emissions from a full-scale deammonification plant. *Water Environment Research*, **82**(2), 169–175.

Weissenbacher N., De Clippeleir H., Boeckx P., Hell M., Chandran K., Murthy S. and Wett B. (2011). Control of N_2O-emissions from sidestream treatment. *Proceedings of the Water Environment Federation*, **2011**(18), 236–248.

Wrage N., Velthof G. L., Oenema O. and Laanbroek H. J. (2004). Acetylene and oxygen as inhibitors of nitrous oxide production in *Nitrosomonas europaea* and *Nitrosospira briensis*: a cautionary tale. *FEMS Microbiology Ecology*, **47**(1), 13–18.

Yang W., Zhao Q., Lu H., Ding Z., Meng L. and Chen G.-H. (2016). Sulfide-driven autotrophic denitrification significantly reduces N_2O emissions. *Australian Journal of Marine and Freshwater Research*, **90**, 176–184.

Yang Y., Chen T., Morrison L., Gerrity S., Collins G., Porca E., Li R. and Zhan X. (2017a). Nanostructured pyrrhotite supports autotrophic denitrification for simultaneous nitrogen and phosphorus removal from secondary effluents. *Chemical Engineering Journal*, **328**, 511–518.

Yang Y., Chen T., Sumona M., Gupta B. S., Sun Y., Hu Z. and Zhan X. (2017b). Utilization of iron sulfides for wastewater treatment: a critical review. *Reviews in Environmental Science and Bio/Technology*, **16**(2), 289–308.

Yu R., Kampschreur M. J., Loosdrecht M. C. v. and Chandran K. (2010). Mechanisms and specific directionality of autotrophic nitrous oxide and nitric oxide generation during transient anoxia. *Advances in Environmental Science and Technology*, **44**(4), 1313–1319.

Zeng R. J., Lemaire R., Yuan Z. and Keller J. (2003). Simultaneous nitrification, denitrification, and phosphorus removal in a lab-scale sequencing batch reactor. *Biotechnology and Bioengineering*, **84**(2), 170–178.

Zhang L., Zhang C., Hu C., Liu H. and Qu J. (2015). Denitrification of groundwater using a sulfur-oxidizing autotrophic denitrifying anaerobic fluidized-bed MBR: performance and bacterial community structure. *Applied Microbiology and Biotechnology*, **99**(6), 2815.

Zhang M., Lawlor P. G., Li J. and Zhan X. (2012). Characteristics of nitrous oxide (N_2O) emissions from intermittently-aerated sequencing batch reactors treating the separated liquid fraction of anaerobically digested pig manure. *Water, Air, & Soil Pollution*, **223**(5), 1973–1981.

Zhang Z., Zhu J., King J. and Li W. (2006). A two-step fed SBR for treating swine manure. *Process Biochemistry*, **41**(4), 892–900.

Zheng H., Hanaki K. and Matsuo T. (1994). Production of nitrous oxide gas during nitrification of wastewater. *Water Science and Technology*, **30**(6), 133–141.

Zhou Y., Pijuan M., Zeng R. J. and Yuan Z. (2008). Free nitrous acid inhibition on nitrous oxide reduction by a denitrifying-enhanced biological phosphorus removal sludge. *Advances in Environmental Science and Technology*, **42**(22), 8260–8265.

Zhu X., Chen Y., Chen H., Li X., Peng Y. and Wang S. (2013). Minimizing nitrous oxide in biological nutrient removal from municipal wastewater by controlling copper ion concentrations. *Applied Microbiology and Biotechnology*, **97**(3), 1325–1334.

Chapter 7

Life cycle assessment of a wastewater treatment plant

William Finnegan, Guangxue Wu and Xinmin Zhan

7.1 DIRECT AND INDIRECT GHG EMISSIONS

Wastewater treatment has become an important issue due to the pollution associated with the emission of untreated or poorly treated wastewater into waterways. Consequently, large financial investments have been stimulated with the aim of installing new wastewater treatment plants (WWTPs) to reduce the concentrations of contaminants in wastewaters, ensure the mitigation of pollution effects and avoid damage to the environment and ecology. The sustainability associated with wastewater treatment has become a growing concern globally in recent years. There are three main aspects, which need to be considered, when establishing the sustainability of any product or service: economic, social and environmental impacts. A number of different assessment tools may be used to estimate sustainability, such as exergy analysis, economic analysis and life cycle assessment (LCA). LCA is a systematic technique that may be applied to a product or service, in this case a WWTP, to estimate the environmental impacts associated with its inputs and emissions.

In recent years a large number of LCAs have been performed relating to wastewater treatment technologies, varying from large scale (Hospido *et al.* 2008) to small scale (Bravo & Ferrer, 2011) wastewater treatment plants, with differences in wastewater treatment technologies (Kalbar *et al.* 2013; Li *et al.* 2013b). LCA has become a widely used tool for evaluation of the sustainability of WWTPs and assessment of the greenhouse gas (GHG) emissions during the construction and operation.

In this chapter, the use of LCA, including background and a brief look at the history since its inception, is introduced. Also, the stages involved in performing an LCA according to the International Organization for Standardization (ISO)

Xinmin Zhan, Zhenhu Hu, Guangxue Wu
doi: 10.2166/9781780406312_127

standards, primarily ISO 14040 (ISO, 2006a) and ISO 14044 (ISO, 2006b), are summarized and discussed. The implementation of LCA to examine the environmental impacts of a WWTP is explored. Additionally, a case study example of an LCA is performed on a municipal WWTP in Kunming City, China in order to establish the environmental impacts of the plant.

7.2 LIFE CYCLE ASSESSMENT

LCA is a method which is used to assess the environmental aspects and potential impacts associated with all the stages of a product's life cycle. This technique applies a systematic set of procedures for compiling and examining an inventory of relevant energy usage, material inputs, environmental emissions and outputs. There are two main approaches when performing an LCA, which are attributional and consequential LCAs. The attributional LCA uses average data, which represent the average environmental burden of the activity, while the consequential LCA uses marginal data, which represent the effect of a small change within the activity. The main advantage of LCA is its ability to account for both direct and indirect impacts related to a product or service throughout its life cycle, for example the inclusion of the manufacturing of input resources and the generation of electricity consumed.

LCA allows the identification of the stages within the life cycle which cause the greatest impact on the environment and identifies the predominant impact. The results of the assessment can, therefore, be used to increase resource usage efficiencies while, also, decreasing liabilities. As a result, one of the main applications of LCA is as a kind of decision-making tool as it provides a comprehensive overview of the environmental effects of a product over its entire life span. LCA can also be used to identify improvement opportunities relating to the environmental performance of a product and used in marketing of a product.

LCA is also known as cradle-to-grave analysis as the process examines all the stages of a product's life cycle from the "cradle" to the "grave" (i.e., raw material extraction, material processing, product manufacture, operational use and maintenance, and end-of-life disposal or recycling). An example of the life cycle of a product from cradle-to-grave is shown in Figure 7.1.

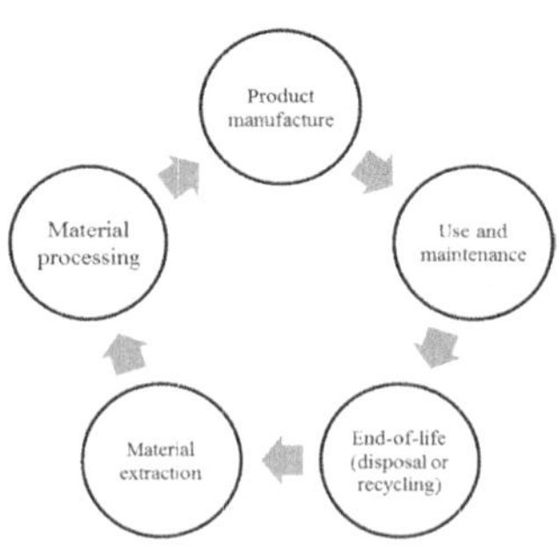

Figure 7.1 Life cycle of a product from cradle to grave.

7.3 A BRIEF HISTORY OF LIFE CYCLE ASSESSMENT

LCA had its beginnings in the 1960s with the first LCA study being commissioned by Coca-Cola in 1969. In this study, the Coca-Cola Company quantified the impacts of different beverage containers on the environment. This study determined which container had the lowest environmental impact and least effect on the supply of natural resources. However, LCA did not enter the public domain until 1974 when the United States Environmental Protection Agency (USEPA) published a report entitled: "Resource and Environmental Profile Analysis of Nine Beverage Container Alternatives" (Hunt *et al.* 1974).

In 1975, due to the oil crisis, there was a shift of focus from waste production and emissions to energy requirements and the energy portion of the LCA (Hunt & Franklin, 1996). Furthermore, European interest in LCA increased with the establishment of the Environmental Directorate (DG X1) by the European Commission (Scientific Applications International Corporation [SAIC], 2006). The DG X1 issued the Liquid Food Container Directive in 1985, which challenged member companies with monitoring their energy inputs, raw material consumption and solid waste generation.

With solid waste as a worldwide issue in 1988, LCA was implemented again as an instrumental tool for analyzing environmental problems. In the 1990s, concerns relating to the inappropriate use of LCA led to the development of standardized LCA methodologies in the ISO 14000 series (SAIC, 2006). The most applicable standards today are: ISO 14040 Environmental Management – Life Cycle Assessment – Principles and Framework (ISO, 2006a) and ISO 14044 Environmental Management – Life Cycle Assessment – Requirements and Guidelines (ISO, 2006b).

With regard to wastewater treatment, the first LCA study was published in 1995 by Emmerson *et al.* (1995). The study examined a number of small-scale wastewater treatment technologies and was mainly focused on the inventory phase.

7.4 STAGES OF LIFE CYCLE ASSESSMENT

ISO LCA standards define the framework and methodology, which are outlined in ISO 14040 (ISO, 2006a), necessary to conduct a valid credited international LCA study. The four ISO standardized practice phases, which are summarized in Figure 7.2 (Source: ISO 14040 (ISO, 2006a)), are: (1) goal and scope definition; (2) life cycle inventory analysis; (3) impact assessment; and (4) interpretation of results. Additionally, when performing an LCA, the requirements of ISO 14044 (ISO, 2006b) shall apply.

7.4.1 Goal and scope definition

Any LCA study begins with the definition of the goal and scope of the study. Identification of the proposed application, intended audience, reasons for carrying out the study, functional unit, product system, data quality requirements, study limitations, boundaries and assumptions are documented in this first LCA stage. In order to make efficient and effective use of time and resources, the following steps are carried out:

- Definition of the goal(s) and main objectives of the project, including the intended application and audience;
- Determination of the information required by decision makers;
- Determination of the scale of the study (e.g., generic study or product specific);
- Definition of a functional unit to relate the data and results to;
- Definition of the scope of the study (i.e., system boundaries); and
- Determination of the assumptions, limitations, level of detail and reporting requirements of the study.

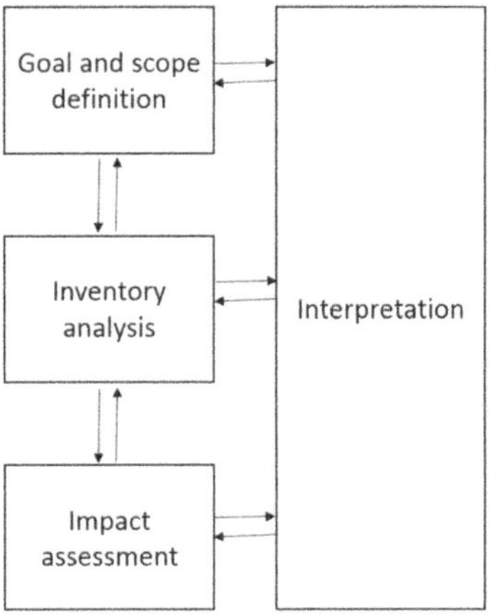

Figure 7.2 Stages of the LCA framework (Adapted from: ISO 14040 (ISO, 2006a)).

7.4.2 Life cycle inventory analysis

The life cycle inventory (LCI) stage creates an inventory of the nature of flows from and to the product or system (ISO, 2006a). This is the most significant LCA study stage as the quality of the data plays an important role in determining the credibility of the study. The primary phase objectives include data collection, data calculation and the allocation of inputs and outputs according to system boundaries (Ramesh *et al.* 2010). Environmental LCI calculation methodologies are categorized into two distinct types, namely: process-based (PB) and input-output (I-O) LCA models. The PB LCA model itemizes the inputs (materials, energy resources, etc.) and the outputs (emissions to air, water and soil, and discharge of wastes to the environment, etc.) associated with a given step in the manufacture of a product. The I-O LCA model uses the history of economic transactions along a given supply chain in order to evaluate where value has been added or removed. A holistic LCA approach incorporating features of both types is referred to as a hybrid LCA methodology. Variations exist in all methodology types with various advantages and disadvantages attached to each (Treloar, 1998; Crawford, 2005; McCaffrey, 2011).

The LCI is a process of collection and quantification of the energy, raw materials, emissions, water/solid waste and other releases that come from the entire life cycle of a product or service (ISO, 2006a). In this process, it is necessary to determine the inputs and emissions involved in a process. An example of the inputs, extractions and emissions associated with a process is given in Figure 7.3.

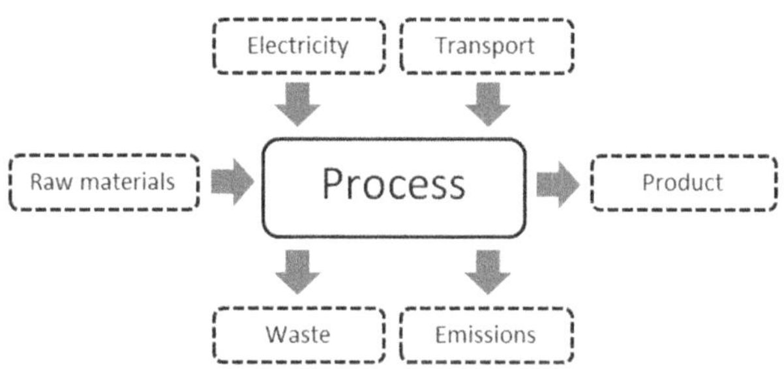

Figure 7.3 Inputs, extractions and emissions associated with a process.

Once an inventory of the inputs and emissions associated with a product is compiled, it is necessary to determine the elementary flows associated with these. This is commonly determined by the life cycle inventory databases (OpenLCA Nexus, 2017). Well known examples of these databases are ecoinvent (Weidema *et al.* 2013) and European reference Life Cycle Database (ELCD, 2012). When performing a life cycle inventory analysis, the next stage is the interpretation of the results.

7.4.3 Life cycle impact assessment

The life cycle impact assessment (LCIA) evaluates the environmental impacts associated with the LCI results, within the system boundaries. The process applies a number of environmental impact categories, which are consistent with the goal and scope of the study, to the inventory data, while providing information for the life cycle interpretation phase. Eleven environmental impact categories are widely used, among which global warming potential is one. Units and definitions of these environmental impact categories used in LCA studies are presented in Table 7.1.

Table 7.1 Summary of commonly used impact categories (Adapted from: Acero *et al.* (2015)).

Impact Category	Unit	Definition
Acidification	kg SO_2 eq.	Reduction of the pH due to the acidifying effects of anthropogenic emissions
Global warming potential	kg CO_2 eq.	Alteration of global temperature caused by greenhouse gases
Depletion of abiotic resources	kg antimony eq. kg minerals MJ fossil fuels m^3 water consumption	Decrease of the availability of non-biological resources (non- and renewable) as a result of their unsustainable use
Eco-toxicity	kg 1,4-DB eq.	Toxic effects of chemicals on an ecosystem
Eutrophication	kg PO_4^{3-} eq. kg N eq.	Accumulation of nutrients in aquatic systems
Human toxicity	kg 1,4-DB eq.	Toxic effects of chemicals on humans
Ionizing radiation	kg U^{235} eq.	Type of radiation composed of particles with enough energy to liberate an electron from an atom or molecule
Land use	m^2a	Impact on the land due to agriculture, anthropogenic settlement and resource extractions
Ozone layer depletion	kg CFC-11 eq.	Diminution of the stratospheric ozone layer due to anthropogenic emissions of ozone depleting substances
Particulate matter	kg particulate matter	Suspended extremely small particles originated from anthropogenic processes such as combustion, resource extraction, etc.
Photochemical oxidation	kg ethylene eq. kg NMVOC kg formed ozone	Type of smog created from the effect of sunlight, heat and NMVOC and NOx

Note: 1,4-dichlorobenzene (1,4-DB); carbon dioxide (CO_2); chlorofluorocarbon-11 (CFC-11); nitrogen (N); non-methane volatile organic compounds (NMVOC); phosphate (PO_4^{3-}); sulfur dioxide (SO_2); square metre-years (m^2a); uranium-235 (U^{235}).

The mandatory elements of the LCIA phase, according to ISO 14040 (ISO, 2006a), are:

1. Selection of environmental impact categories, category indicators and characterization methods;

2. Classification: assignment of inventory data (LCI results);
3. Characterization: calculation of category indicator results; and
4. LCIA profile: evaluating and reporting of the category indicator results.

Optional elements and their definitions, according to ISO 14040 (ISO, 2006a), of the LCIA, are as follows:

- normalization: calculating the magnitude of category indicator results relative to reference information;
- grouping: sorting and possibly ranking of the impact categories; and
- weighting: converting and possibly aggregating indicator results across impact categories using numerical factors based on value-choices; data prior to weighting should remain available.

However, a limitation of LCIA is that only the environmental impacts that are detailed in the goal and scope stage of the study are assessed. Therefore, the analysis is not a complete assessment of all environmental impacts of the product.

7.4.4 Interpretation

The final phase within the LCA is the interpretation. In this stage, the results from the LCI and the LCIA are discussed and evaluated. The system functions, the functional unit and system boundary, which are defined in the goal and scope stage of the study, along with any limitations identified are discussed and reviewed. Suggestions for improvements to the life cycle of the product, which are consistent with the goal and scope of the study, are also included in the interpretation.

There are three main elements in conducting the interpretation phase, according to ISO 14044 (2006b):

- The identification of the significant issues based on the LCI and the LCIA;
- An evaluation which includes completeness, sensitivity and consistency checks; and
- Conclusions, recommendations and reporting.

Additionally, the interpretation may take the form of conclusions and recommendations that are reported to decision makers, provided that they are consistent with the goal and scope of the study (ISO, 2006b).

7.5 LCA OF A WWTP

To illustrate the effectiveness of LCA, an example of a process-based attributional LCA is performed in order to assess the environmental impact of a WWTP. The initial stage of the LCA of a WWTP is to establish the scope and goals of the study. The system boundary and the life cycle stages, which are to be included, need to be established as this will have a significant impact on the results. Generally, the system only includes the plant itself, and the transportation of wastewater (through pipelines, sewer system, trucks, etc) to the plant is not included. However, waste management, which includes sludge treatment and disposal, may be included as a downstream impact of the WWTP.

The main life cycle stages for a WWTP, which are illustrated in Figure 7.4, are:

- Production of raw materials, which includes extraction and processing;
- Construction of the plant;
- Operation and maintenance of the plant; and
- End-of-life stages: demolition, recycling of materials and elements, landfilling etc.

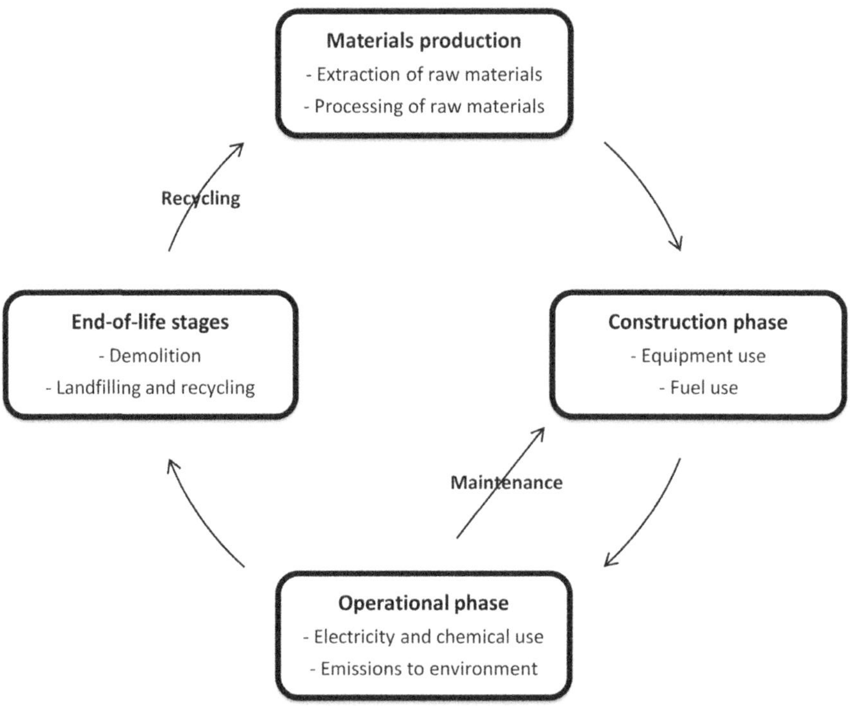

Figure 7.4 Life cycle of a WWTP from cradle to grave.

When performing an LCA, the impact associated with some inputs, emissions or, even, stages may prove to be insignificant. However, it is necessary to ensure that ISO guidelines are followed when omitting aspects within an LCA. A number of previous studies on LCA of WWTPs have demonstrated that the construction and demolition phases of the life cycle have negligible impacts compared to the operation phase due to the short duration of the construction stage, compared with the long operation span of the WWTPs (Tillman *et al.* 1998; Renou *et al.* 2008; Li *et al.* 2013b): the GHG emission ($CO_{2\text{-eq}}$) during the construction stage is only 11% of the GHG emissions ($CO_{2\text{-eq}}$) during the operation phase (Renou *et al.* 2008), and the operation and maintenance phase would contribute 99.5% of the environmental impacts (Li *et al.* 2013b). Therefore, an LCA which deals with the operation phase of a WWTP's life cycle (i.e., a gate-to-gate analysis of the WWTP) will provide an accurate representation of the environmental impacts of the plant over its "whole" life cycle. However, for wetland wastewater treatment systems the construction phase should be taken into account; for membrane bioreactors the construction phase may also be significant (Corominas *et al.* 2013). Nevertheless, if sufficient data is available for the construction phase of the life cycle (i.e., quantities of steel, concrete and other materials), they may be included in the analysis and, thereby, increase the accuracy of the analysis.

An important aspect when establishing the goals of the study is selecting a suitable functional unit. LCA of a WWTP presents a number of options available regarding the functional unit. The two most commonly used functional units are the volume of wastewater treated, for example per cubic metre (m^3), and the treatment of wastewater generated from one-person equivalent (p.e.), which is defined as sewage flow containing 60 g of 5-day biochemical oxygen demand (BOD_5) load per day (Corominas *et al.* 2013). The functional unit of the study may also include a time span, for example, the life span of the WWTP. Yet, in some cases it may be more suitable to use other functional units, for example, per kg of chemical oxygen demand (COD) removed when analyzing the treatment of wastewater from dairy processing plants. Largely, when analyzing WWTPs, the volume of wastewater treated is, generally, the most suitable and widely used option.

When compiling the life cycle inventory, for the operational phase of a WWTP, a number of aspects should be included:

- Resource consumption: electricity, chemicals, etc.;
- Influent wastewater into plant;
- Emissions to water source (plant effluent);
- Disposal of biosolids; and
- Emissions to the atmosphere (during plant operation).

Previously, electricity consumption within the WWTP was identified as the dominant contributor to environmental impacts analyzed (Gallego *et al.* 2008). This is the case as electricity is required at almost every stage within the operation of a WWTP, including pumping, mixing, heating and aeration. However, the technologies used in a WWTP can greatly influence its environmental impact. For instance, an activated sludge system requires more energy than either a lagoon or land treatment systems (Muga & Mihelcic, 2008). Additionally, the mode of electricity generation used is a source of possible discrepancy between assessments conducted in different countries, or a significant number of years apart, and this is particularly important to account for when performing comparisons.

As described in the previous chapters, the three main emissions to the atmosphere during the operation of a WWTP are CO_2, CH_4 and N_2O. CO_2 emissions from wastewater are not considered as these are of biogenic origin. It is difficult to quantify these emissions but a number of studies have been performed which estimate the direct CH_4 emissions (El-Fadel & Massoud, 2001; Wang *et al.* 2011a; Daelman *et al.* 2012b) and direct N_2O emissions (Kampschreur *et al.* 2009c; Foley *et al.* 2010; Hu *et al.* 2011b) from the treatment of municipal wastewater.

There are a significant number of environmental impact assessment methods available. The most widely used impact assessment methods in wastewater treatment LCA are CML 2000, Eco Indicator 99, EDIP 96, EPS and Ecopoints 97, and the choice of method is not a critical issue when examining wastewater treatment as the results they provide are similar (Renou *et al.* 2008). However, when assessing WWTPs, a number of previously published studies (Gallego *et al.* 2008; Hospido *et al.* 2008; Bravo & Ferrer, 2011; Kalbar *et al.* 2013; Li *et al.* 2013b) have used the CML (Baseline) 2000 method. Additionally, normalization of impact factors may be performed in order to compare the environmental impacts on the same scale.

The final stage of any LCA is the interpretation, as described in Section 7.7.4.4, and this stage may also include a sensitivity analysis in order to determine which parameters influence the LCA outcome most.

7.6 CASE STUDY: LCA OF A MUNICIPAL WWTP IN KUNMING CITY, CHINA

7.6.1 Introduction

In order to demonstrate the advantages and the insight provided by an environmental LCA when applied to a full-scale municipal WWTP, a case study example of a municipal WWTP in Kunming City, China is presented.

The WWTP is a full-scale municipal WWTP with a treatment capacity of approximately 300,000 m^3/day. The unit processes used in the plant are given in Figure 7.5. The biological treatment process within the plant is anoxic/anaerobic/aerobic (A^2O). The effluent water is also passed through a tertiary treatment system, which involves filtration and ultraviolet disinfection, before being discharged to the receiving river. The separated sludge treatment consists of sludge thickening and dewatering, followed by transportation to landfill for disposal.

In this study, the environmental impacts of the municipal WWTP in Kunming, China are assessed through LCA. In order to build up an accurate representation of the plant, two years' data (2011 & 2012) are analyzed and compared. The life cycle inventory is presented for the WWTP for its operational phase. Furthermore, the results are presented for two functional units (volume and person equivalent) and the contribution of the resources, inputs and emissions to each environmental impact are displayed.

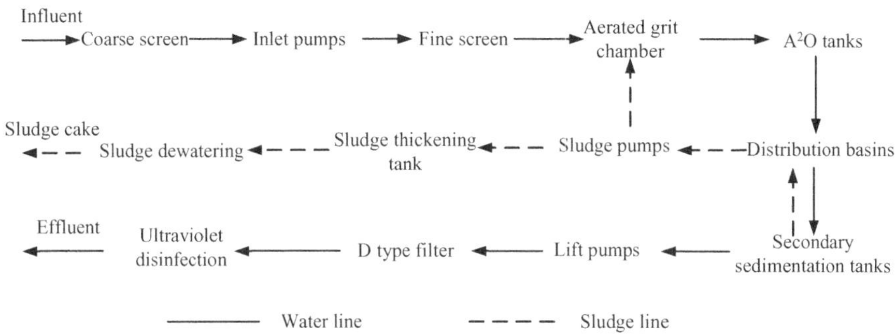

Figure 7.5 Flow diagram of the unit processes within the municipal WWTP in Kunming City, China.

7.6.2 Materials and methods

This study has been structured in accordance with the life cycle assessment guidelines of ISO 14040 (ISO, 2006a) and ISO 14044 (ISO, 2006b). Life cycle inventory data sets based on the ecoinvent LCI database version 3 (Weidema *et al.* 2013) were used in order to account for upstream impacts of input resources (i.e., electricity, chemicals, etc.).

7.6.2.1 Goal and scope definition

In this case, the scope is a gate-to-gate analysis using annual data obtained from 2011 and 2012. A graphical summary of the scope and system boundaries for the LCA study is presented in Figure 7.6. In this study, a gate-to-gate analysis of the WWTP, which accounts for the operation phase of the WWTP's life cycle, is performed as this will provide an accurate representation of the environmental impacts of the plant over its "whole" life cycle. This assumption is discussed in greater detail in Section 7.7.5. Two functional units will be used, in this study, within the case study: per m^3 and per p.e.

7.6.2.2 Life cycle inventory analysis

The life cycle inventory is generated on an annual basis for the municipal WWTP in Kunming City, China. Electricity usage, water consumption, chemical usage, operational emissions, effluent emissions and sludge disposal are included in the life cycle inventory for 2011 and 2012. A summary of the life cycle inventory for the operational phase of the WWTP for 2011 and 2012 is given in Table 7.2.

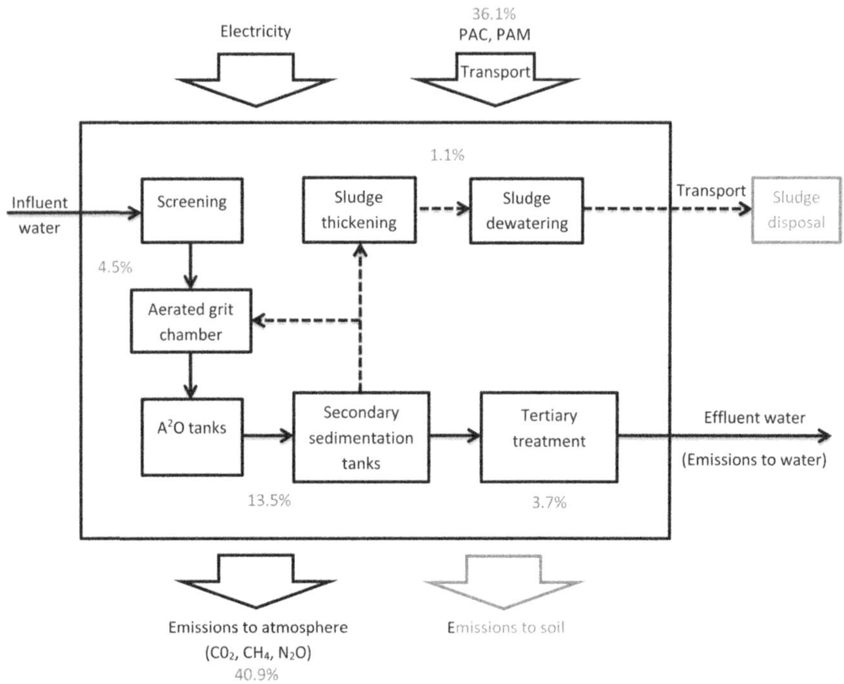

Figure 7.6 Scope and system boundaries for the LCA study. Gray features have been omitted from the study. The percentage contributions of the main contributors to GWP, based on this study, have been indicated in red text. PAC: poly aluminum chloride; PAM: polyacrylamide.

Electricity

The total electricity used for the operation of the WWTP (including wastewater treatment, sludge treatment, utilities etc.) was recorded on an annual basis by plant operators. The emissions factors for the electricity used were obtained from the ecoinvent LCI dataset (Weidema *et al.* 2013) for the electricity generation mix of China. However, the global warming potential of electricity produced in the South China region is reported to be 0.3791 kg $CO_{2\text{-eq}}$ kWh^{-1} (CCChina, 2013) and this value has been used in the analysis.

A breakdown of the energy usage within the plant is given in Table 7.3. The majority of the energy usage occurred in the pre-treatment and secondary treatment processes with air blowers and inlet pumps contributing 36.1% and 18.6% of the total energy consumption, respectively.

Table 7.2 Life cycle inventory for the municipal WWTP in Kunming City, China for 2011 and 2012.

	Unit	2011	2012
Influent:			
Flow	m^3/yr	110,047,766	109,618,333
BOD_5	mg/L	236.4	222.0
COD	mg/L	405.1	382.2
SS	mg/L	419.0	394.8
NH_4^+-N	mg/L	21.4	22.2
TN	mg/L	37.5	38.1
TP	mg/L	8.0	7.8
Effluent:			
BOD_5	mg/L	1.7	1.2
COD	mg/L	22.2	22.8
SS	mg/L	5.2	5.3
NH_4^+-N	mg/L	0.5	0.6
TN	mg/L	12.6	11.5
TP	mg/L	0.4	0.3
HRT[a]	hour	18	18
Yield of sludge	kg/yr	63,012	81,833
Sludge Transportation	km	32	32
Electricity consumption	kWh/yr	22,968,534	23,864,597
Water consumption	m^3/yr	13,839	1,857
PAC	kg/yr	4,047,343	4,979,260
PAM	kg/yr	31,415	45,815
Chemical transportation	km	60	60

[a] HRT: hydraulic retention time.

Chemicals

The two main chemicals used at the WWTP are poly aluminum chloride (PAC), which is used as a coagulant in the removal of phosphorous from the wastewater, and polyacrylamide (PAM), which is used as a flocculant in the sludge conditioning process. Only data for PAM was available in the ecoinvent LCI database version 3 (Weidema *et al.* 2013) and, therefore, this was used to model both PAC and PAM for the purpose of this study, which was similar to De Haas *et al.* (2008). The transportation of chemicals to the WWTP is also included in the study, where the transportation distance is assumed to be 60 km.

Table 7.3 Energy consumption of equipment in the municipal WWTP in Kunming City, China.

Processing Unit	Equipment	Installed Power (kW)	Theoretical Energy Consumption (kWh/day)	Percentage within Each Processing Unit (%)	Percentage of Total (%)
Pre-treatment Process	Coarse screen	8.85	70.8	0.4	0.1
	Thin screen	18	144	0.9	0.2
	Inlet pump	2220	15480	94.6	18.7
	Roots blower	115	600	3.7	0.7
	Sand-suction machine	3	72	0.4	0.1
Secondary treatment	Air blowers	2220	29840	61.4	36.1
	Mixers	225	5400	11.1	6.5
	Mixed liquid recycle pump	450	7200	14.8	8.7
	Sludge recycle pump	423	1944	4.0	2.3
	Excess sludge pump	333	2664	5.5	3.2
	Aspiration sludge scraper	66	1584	3.3	1.9
Sludge treatment	Sludge feed pump	75	450	10.6	0.5
	Helical rotor pump	100	600	14.1	0.7
	Sludge cutting machine	17.5	165	3.9	0.2
	Sludge inlet pump	75	450	10.6	0.5
	Centrifugal dehydrator	430	2580	60.8	3.1
Tertiary treatment	UV disinfection tank	256	6144	45.5	7.4
	Reclaimed water lift pump	44	528	3.9	0.6

(*Continued*)

Table 7.3 Energy consumption of equipment in the municipal WWTP in Kunming City, China (*Continued*).

Processing Unit	Equipment	Installed Power (kW)	Theoretical Energy Consumption (kWh/day)	Percentage within Each Processing Unit (%)	Percentage of Total (%)
	Backwash blower	60	270	2.0	0.3
	Backwash pump	88	396	2.9	0.5
	Dosing pump	7.50	180	1.3	0.2
	Tertiary treatment lift pump	625	6000	44.4	7.2
Total			82761.8		100

WWTP operation

In this analysis, similar to Li *et al.* (2013b), the direct methane emissions were calculated using the emission factor defined by Wang *et al.* (2011a). Wang *et al.* (2011a) calculated the direct methane emissions from the WWTP they investigated, which also used an A^2O process, to be 1.55×10^{-4} g CH_4 per litre of wastewater treated. Additionally, the direct nitrous oxide emissions were calculated using 0.018 kg N_2O-N (kg $N_{denitrified}$)$^{-1}$, which is the nitrous oxide generation factor observed by Foley *et al.* (2010) for an A^2O process.

Effluent to water source

The emissions to water, in this case a nearby river, caused by nutrients in the effluent waters are accounted for in this study. The nutrients which are included are given in Table 7.2 and include biological oxygen demand (BOD_5), chemical oxygen demand (COD), suspended solids (SS), ammonia (NH_4^+-N), total nitrogen (TN) and total phosphorous (TP). Heavy metals emission within effluent waters has been omitted from the study as there were not sufficient data available from the plant.

Disposal of biosolids

The biosolids, or sludge, from the WWTP is disposed of at landfill. The transportation of sludge to landfill is included in the study and the distance was reported to be, on average, 32 km. However, the method of disposal of the sludge, including the resultant emissions to soil, has been omitted from the study.

7.6.2.3 Impact assessment

Similar to a number of previously published studies (Gallego *et al.* 2008; Hospido *et al.* 2008; Bravo & Ferrer, 2011; Kalbar *et al.* 2013; Li *et al.* 2013b), the method for assessing the environmental impact used in this study is the CML (Baseline) 2000 method. This method calculates the following environmental impacts:

- Abiotic depletion (AD)
- Acidification (AP)
- Eco-toxicity (ET)
- Eutrophication (EP)
- Global warming potential (GWP)
- Human toxicity (HT)
- Ozone layer depletion (ODP)
- Photochemical oxidation (PCO)

Detailed definitions and descriptions of these environmental impacts are given in Table 7.4.

Table 7.4 Results of the life cycle impact assessment of the municipal WWTP in Kunming City, China.

Environmental Impact	Unit	2011		2012	
		per m^3	per p.e.	per m^3	per p.e.
Abiotic depletion	kg Sb eq.	3.87×10^{-7}	3.61×10^{-5}	4.77×10^{-7}	4.73×10^{-5}
Acidification	kg SO_2 eq.	2.93×10^{-3}	0.274	3.16×10^{-3}	0.314
Eco-toxicity	kg 1,4-DB eq.	330	30840	357	35363
Eutrophication	kg PO_4^{3}- eq.	7.82×10^{-3}	0.730	7.07×10^{-3}	0.701
Global warming potential	kg CO_2 eq.	0.322	30	0.359	35.6
Human toxicity	kg 1,4-DB eq.	8.66×10^{-2}	8.08	9.69×10^{-2}	9.61
Ozone layer depletion	kg CFC-11 eq.	7.7×10^{-9}	7.19×10^{-7}	9.26×10^{-9}	9.19×10^{-7}
Photochemical oxidation	kg C_2H_4 eq.	1.12×10^{-4}	0.01	1.21×10^{-4}	0.012

7.7 INTERPRETATION OF RESULTS AND DISCUSSION

The results of the life cycle impact assessment of the municipal WWTP are displayed in Table 7.4 and a baseline comparison of the two years (using the 2011 results as the baseline, 100%) is displayed graphically in Figure 7.7 for the two functional units (per m^3 and per p.e.). The p.e. values for 2011 and 2012 were found to be 1,179,299 and 1,105,462, respectively. Between 2011 and 2012, the environmental impact of the WWTP has increased, per functional unit, across most of the impact categories except for eutrophication, which is evident from Figure 7.7. The decrease in eutrophication is attributed to the reduced concentration of total nitrogen and total phosphorous in the plant effluent reported in 2012 compared to 2011. This is the case as the emission from the WWTP to the water course is the main contributor to eutrophication, which can be seen from Figure 7.8.

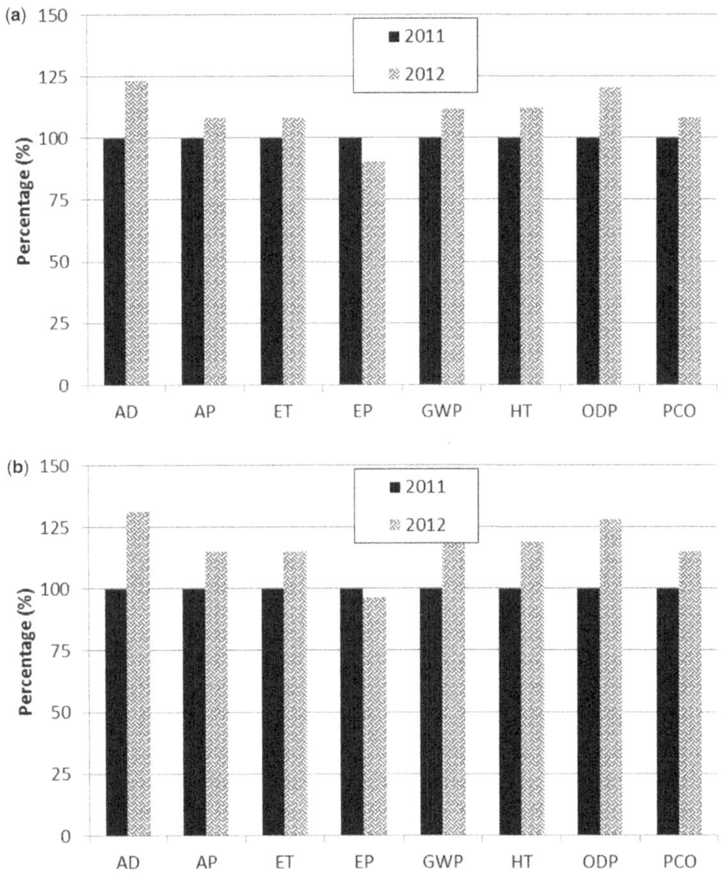

Figure 7.7 Baseline percentage comparison of the environmental impacts on the municipal WWTP in Kunming City, China between 2011 and 2012 (a): per m^3 and (b): per p.e. *Note*: Abiotic depletion (AD); Acidification (AP); Eco-toxicity (ET); Eutrophication (EP); Global warming potential (GWP); Human toxicity (HT); Ozone later depletion (ODP); Photochemical oxidation (PCO).

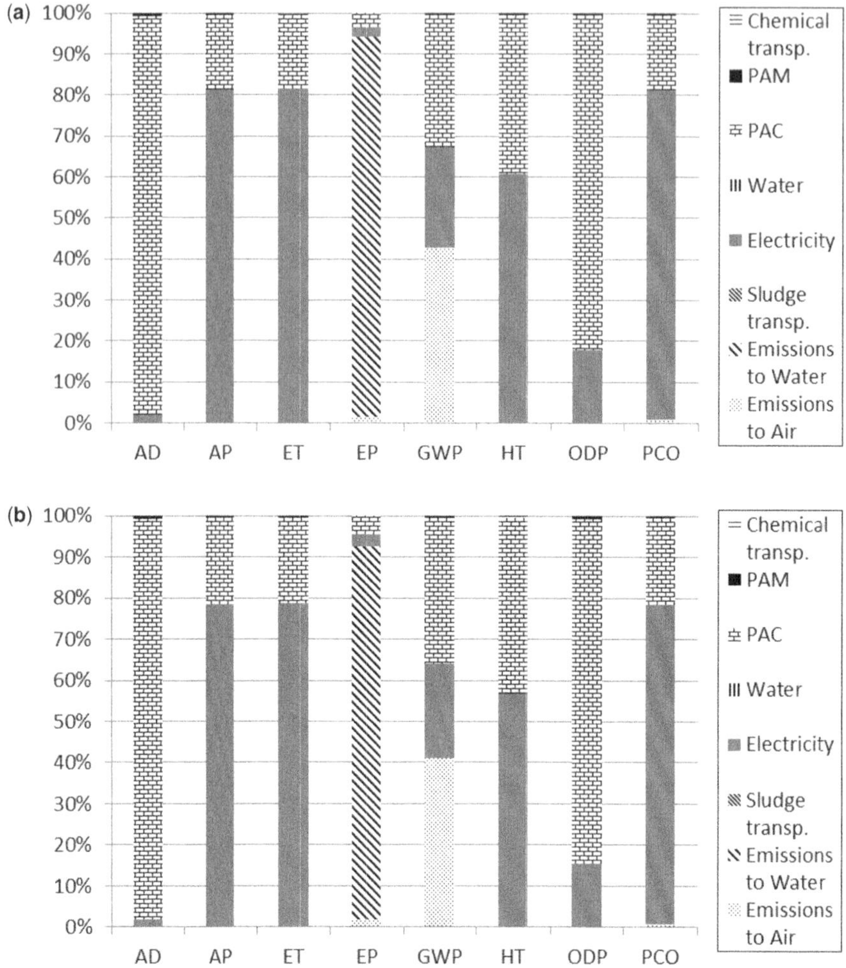

Figure 7.8 Contribution of the resources, inputs and emissions to each environmental impact per m^3 for (a): 2011 and (b): 2012. Note: Abiotic depletion (AD); Acidification (AP); Eco-toxicity (ET); Eutrophication (EP); Global warming potential (GWP); Human toxicity (HT); Ozone later depletion (ODP); Photochemical oxidation (PCO).

The contribution of the resources, inputs and emissions to each environmental impact per m^3 is shown in Figure 7.8. It is clear to see that the most significant contributor varies between a number of inputs and emissions, dependent on the impact category. For example, abiotic depletion (AD) and ozone layer depletion (ODP) are mainly affected by the quantity of chemicals used and eutrophication (EP) is mainly affected by the effluent emissions to the water course – from the wastewater treatment plant. Additionally, electricity is the most significant contributor to acidification (AP), eco-toxicity (ET) and photochemical oxidation (PCO) at approximately 78–81% of the total impact of each category. However, electricity, emissions to air (in the form of CH_4 and N_2O) and PAC are major contributors to global warming potential (GWP).

Similar contributions from inputs and emissions occurred in 2011 and 2012, which can be seen in Figures 7.8 (a) and (b), respectively.

From the LCA, the main contributors to the environmental impacts of the WWTP were identified and found to be electricity consumption, PAC usage and direct emissions to water and air. Between 2011 and 2012, an increase in the environmental impact of the WWTP, per functional unit, has been observed across all the impact categories, except for eutrophication. One of the main contributors to the overall increase in environmental impacts is the increase in electricity consumption and usage of chemicals, which were required to reduce the direct emissions to water between 2011 and 2012. This is an example of how LCA may provide decision makers with information when managing the operation of the plant. The LCA results clearly show that the improvement in effluent quality was very limited in 2012 compared in 2011. Therefore, there should be a balance considered between treated water quality and indirect emissions.

A management option for reducing the environmental impact of the WWTP is the production of electricity from biogas generated during the treatment processes. Biogas can be produced at a rate of 0.11 m^3 biogas/m^3 wastewater treated, which can be converted into 0.68 kWh/m^3 biogas of electricity (Pasqualino *et al.* 2009). Therefore, suppose an anaerobic digestion system was installed at the municipal WWTP, the amount of electricity produced would supplement 34–36% of the electricity currently consumed. This, in turn, would reduce acidification (AP) by approximately 29%, eco-toxicity (ET) by approximately 29%, global warming potential (GWP) by approximately 9% and photochemical oxidation (PCO) by approximately 28%, as electricity is the most significant contributor to these impacts.

Additionally, the results of this analysis may be used in future decision-making relating to resource usage within the plant. For example, reduction of the use of PAC would have a significant beneficial effect on a number of impact categories. Similarly, the use of biosolids for on-site electricity generation would supplement the use of electricity, while also benefiting the effects of sludge disposal and reducing operational costs.

7.8 SUMMARY

Sustainability within WWTPs will be an ever-existing issue as WWTPs play a vital role in ensuring high water environment quality, which is essential for life and the protection of human health. However, in addition, the WWTP needs to operate in a manner which reduces consumption of resources and energy, reduces waste generation, reduces greenhouse gas emissions and enables the recycling of nutrients.

LCA is a useful tool for quantifying the environmental impacts of a WWTP. LCA explores the impacts associated with the entire system rather than focusing on a single issue within a product's or service's life cycle. Therefore, significance of indirect emissions compared to the direct emissions from the plant can be quantified and used within management decisions.

The reduction in effluent pollutants (mainly nitrogen and phosphorous) and minimizing the production of sludge could be incorporated into the LCA analysis as these are key components when improving the operational efficiency of a WWTP. Furthermore, the boundary of an LCA study may be expanded to include reclaimed water from the system (Pasqualino *et al.* 2011), which would have increased applicability in water-scarce regions.

However, there are a number of drawbacks associated with the method. Often economics is not included in the LCA analysis and, therefore, conclusions made on the results of the LCA may not be reasonably practical. Additionally, risks and safety issues are not quantified within the analysis. The use of other suitable methods and analysis tools, along with LCA, will help to create a better understanding of the effects of WWTPs and aid in building a more efficient and sustainable industry for future generations.

7.9 REFERENCES

Acero A. P., Rodríguez C. and Ciroth A. (2015). LCIA Methods: Impact Assessment Methods in Life Cycle Assessment and Their Impact Categories. Greendelta, Berlin, Germany.

Bravo L. and Ferrer I. (2011). Life Cycle Assessment of an intensive sewage treatment plant in Barcelona (Spain) with focus on energy aspects. *Water Science & Technology*, **64**(2), 440–447.

CCChina (2013). 2012 Baseline Emission Factors for Regional Power Grids in China. China's Department of Climate change, National Development and Reform Commission, Beijing, China.

Corominas L., Foley J., Guest J. S., Hospido A., Larsen H. F., Morera S. and Shaw A. (2013). Life cycle assessment applied to wastewater treatment: state of the art. *Australian Journal of Marine and Freshwater Research*, **47**(15), 5480–5492.

Crawford R. H. (2005). Validation of the use of input-output data for embodied energy analysis of the Australian construction industry. *Journal of Construction Research*, **6**(1), 71–90.

Daelman M. R. J., van Voorthuizen E. M., van Dongen U. G. J. M., Volcke E. I. P. and van Loosdrecht M. C. M. (2012). Methane emission during municipal wastewater treatment. *Australian Journal of Marine and Freshwater Research*, **46**(11), 3657–3670.

De Haas D., Foley J. and Barr K. (2008). Greenhouse gas inventories from WWTPs – the trade-off with nutrient removal. *Proceedings of the Water Environment Federation*, **2008**(6), 264–285.

ELCD (2012). European Commission – Joint Research Centre – Institute for Environment and Sustainability: European Reference Life Cycle Database (ELCD), Version 3.0 ed. European Commission – Joint Research Centre.

El-Fadel M. and Massoud M. (2001). Methane emissions from wastewater management. *Environmental Pollution*, **114**(2), 177–185.

Emmerson R. H. C., Morse G. K., Lester J. N. and Edge D. R. (1995). The life-cycle analysis of small-scale sewage-treatment processes. *Water and Environment Journal*, **9**(3), 317–325.

Foley J., de Haas D., Yuan Z. and Lant P. (2010). Nitrous oxide generation in full-scale biological nutrient removal wastewater treatment plants. *Australian Journal of Marine and Freshwater Research*, **44**(3), 831–844.

Gallego A., Hospido A., Moreira M. T. and Feijoo G. (2008). Environmental performance of wastewater treatment plants for small populations. *Resources, Conservation and Recycling*, **52**(6), 931–940.

Hospido A., Moreira M. and Feijoo G. (2008). A comparison of municipal wastewater treatment plants for big centres of population in Galicia (Spain). *The International Journal of Life Cycle Assessment*, **13**(1), 57–64.

Hu Z., Zhang J., Xie H., Li S., Wang J. and Zhang T. (2011). Effect of anoxic/aerobic phase fraction on N_2O emission in a sequencing batch reactor under low temperature. *Bioresource Technology*, **102**(9), 5486–5491.

Hunt R. G. and Franklin W. E. (1996). LCA – how it came about. *The International Journal of Life Cycle Assessment*, **1**(1), 4–7.

Hunt R. G., Franklin W. E., Welch R. O., Cross J. A. and Woodall A. E. (1974). Resource and Environmental Profile Analysis of Nine Beverage Container Alternatives: Final Report. U. S. Environmental Protection Agency, Washington, DC, USA.

ISO (2006a). ISO 14040: Environmental Management – Life Cycle Assessment – Principles and Framework. International Organization for Standardization, Geneva, Switzerland.

ISO (2006b). ISO 14044: Environmental Management – Life Cycle Assessment – Requirements and Guidelines. International Organization for Standardization, Geneva, Switzerland.

Kalbar P. P., Karmakar S. and Asolekar S. R. (2013). Assessment of wastewater treatment technologies: life cycle approach. *Water and Environment Journal*, **27**(2), 261–268.

Kampschreur M. J., Temmink H., Kleerebezem R., Jetten M. S. M. and van Loosdrecht M. C. M. (2009). Nitrous oxide emission during wastewater treatment. *Australian Journal of Marine and Freshwater Research*, **43**(17), 4093–4103.

Li Y., Luo X., Huang X., Wang D. and Zhang W. (2013). Life cycle assessment of a municipal wastewater treatment plant: a case study in Suzhou, China. *Journal of Cleaner Production*, **57**, 221–227.

McCaffrey M. F. (2011). An I-O Hybrid Methodology for Environmental LCA of Embodied Energy and Carbon in Irish Products and Services – A Study of Reinforced Concrete, College of Engineering and Informatics. National University of Ireland, Galway (NUIG), Galway, Ireland.

Muga H. E. and Mihelcic J. R. (2008). Sustainability of wastewater treatment technologies. *Journal of Environmental Management*, **88**(3), 437–447.

Pasqualino J. C., Meneses M., Abella M. and Castells F. (2009). LCA as a decision support tool for the environmental improvement of the operation of a municipal wastewater treatment plant. *Environmental Science & Technology*, **43**(9), 3300–3307.

Pasqualino J. C., Meneses M. and Castells F. (2011). Life cycle assessment of urban wastewater reclamation and reuse alternatives. *Journal of Industrial Ecology*, **15**(1), 49–63.

Ramesh T., Prakash R. and Shukla K. K. (2010). Life cycle energy analysis of buildings: an overview. *Energy and Buildings*, **42**(10), 1592–1600.

Renou S., Thomas J. S., Aoustin E. and Pons M. N. (2008). Influence of impact assessment methods in wastewater treatment LCA. *Journal of Cleaner Production*, **16**(10), 1098–1105.

SAIC (2006). Life Cycle Assessment: Principles and Practice. Scientific Applications International Corporation for U. S. Environmental Protection Agency, Cincinnati, Ohio, USA.

Tillman A.-M., Svingby M. and Lundström H. (1998). Life cycle assessment of municipal waste water systems. *The International Journal of Life Cycle Assessment*, **3**(3), 145–157.

Treloar G. J. (1998). *A Comprehensive Embodied Energy Analysis Framework, Faculty of Science and Technology*. Deakin University, Geelong, Australia.

Wang J., Zhang J., Xie H., Qi P., Ren Y. and Hu Z. (2011). Methane emissions from a full-scale A/A/O wastewater treatment plant. *Bioresource Technology*, **102**(9), 5479–5485.

Weidema B. P., Bauer C., Hischier R., Mutel C., Nemecek T., Reinhard J., Vadenbo C. O. and Wernet G. (2013). The ecoinvent database: Overview and methodology, Data quality guideline for the ecoinvent database version 3, Zurich, Switzerland.

Index

IWA Publishing's authorised EU representative for General Product Safety Regulations is Diane D'Arras, 15 rue Duret, 75116 Paris, France, e-mail: safety@iwap.co.uk.

Printed and bound by CPI Group (UK) Ltd, Croydon, CR0 4YY
16/06/2026
02140359-0001